Growth and Nutrition in the Horse

Also by David P. Willoughby:

The Super-Athletes
The Empire of Equus

Growth and Nutrition in the Horse

David P. Willoughby

South Brunswick and New York: A. S. Barnes and Company
London: Thomas Yoseloff Ltd

© 1975 by A. S. Barnes and Co., Inc.

A. S. Barnes and Co., Inc.
Cranbury, New Jersey 08512

Thomas Yoseloff Ltd
108 New Bond Street
London W1Y OQX, England

Library of Congress Cataloging in Publication Data

Willoughby, David P
    Growth and nutrition in the horse.

    Bibliography: p.
    Includes index.
    1. Horses—Growth. 2. Horses—Feeding and feeds.
3. Horse breeds. 1. Title.
SF279.W55   632.1′08′52   74-9276
ISBN 0-498-01427-4

Printed in the United States of America

Contents

Introduction

It may safely be said that of all the numerous departments connected with horses, the most poorly studied has been that which deals with growth and development. Particularly is this true of horse breeding in the United States. While hundreds, possibly thousands, of books have been published during recent years on the history of the horse, the major physical characteristics and differences of the principal breeds, the arts of equitation, the training and exhibiting of horses, and so on, not one volume has come out in which, specifically, the actual body measurements of adult horses and growing foals are listed. A few papers, buried in little-known technical journals, are about the extent of the publications in the United States.

In 1892 a book appeared entitled *The Exterior of the Horse*, by the French hippologists Goubaux and Barrier. However, the measurements dealt with therein were almost exclusively linear ones—that is, heights, lengths, and breadths such as would be of use only in drawing a horse or in relating the lengths of its limb bones, head, neck, and trunk one to the other. None of the measurements given in *The Exterior of the Horse* concern the *changes* in size, especially body weight, of the *growing foal* from birth to maturity.

Along with the present unprecedented increase in the use of horses in almost every possible capacity, there is an accompanying increased urge among breeders and owners to raise their foals on a schedule that will insure the best possible mature horses. This procedure requires a dual knowledge of what constitutes optimum growth, and of what kind of nutrition best promotes it. Too, along with optimum nutrition, there must be a background of good inheritance; and the growing foals must be properly cared for in other respects, such as exercise, stabling, pasturing, training (for their particular use), and the prevention and treatment of disease. Only by the breeder's steady attention to all these facets of equine development may colts and fillies be enabled to reach their potential in the show ring, on the race track, or simply as well-performing saddle, carriage, or draft horses.

The purpose of this book is, first, to provide sound and reliable information on what constitutes normal growth, and secondly, to tell what kind of nutrition best brings about such growth. While there is much available information on the foods and feeding of horses, statistics on the growth of the foal are relatively few, especially in the United States and with particular reference to popular breeds. Why such a status should exist is difficult to understand, since in no country are there better opportunities for studying and measuring growing foals of various breeds. Consequently, the statistics to date on such measurings are almost entirely in reference to European breeds, especially those of Germany, Poland, and Hungary. Fortunately, among those statistics are some pertaining to such internationally popular breeds as the Arab, the Thoroughbred, the Belgian draft, and others. From this mass of information, plus some observations and measurings of my own, I am able to present—perhaps for the first time in a publication in English—tables of measurements showing normal growth in such popular yet diverse breeds as Arabs, Thoroughbreds, Standardbreds, Morgans, Appaloosas, Quarter Horses, Shetlands, and Drafters. That reliable information on the relationship of weight to linear size is sadly lacking is shown in a recent article on the nutrition of foals, in which the prescribed body weights are as much as 100 pounds in error!

The procedure followed in this study is, first, to present data on the measurements of horses of various breeds as published in articles on the subject by investigators both in Europe and the United States; and secondly, from this information to deduce general or normal trends of growth applicable to all present-day horse breeds. The term *deduce* is used because of the impossibility of perfectly reconciling the observations or measurements of all investigators. One reason for this is the inherent variability of the animals themselves (even within a given breed, age, and sex) ; and another reason, the differences in the measuring techniques applied by various investigators. Nevertheless, running through all the diverse findings there must be a golden mean, and it is this central tendency that most likely expresses the normal course of a foal's development.

The reader may note that most of the statistical data that I used was published some years ago. This is because this information was consulted by me during my extensive investigation of the entire genus. *Equus* (that is, horses of all kinds, both living and fossil), which study was commenced in the 1940s. Fortunately, the physical development of horses, unlike their records on the racetrack, has altered little if any since the turn of the century. This is evidenced, for example, by the measurements taken by E. E. A. Grange in 1893 and by Simon von Nathusius in 1895 of various breeds both light and heavy, which exhibit practically identical heights and weights to those found in the same breeds today.

It is my hope that the information on equine growth and nutrition presented herein will provide breeders everywhere with a reliable yardstick for guiding and controlling the physical development of horses of all breeds along an optimum course from birth to maturity.

Is not the latter [appearance] that deceiving art which makes us wander up and down and take the things at one time which we repent at another both in our actions and in our choice of things great and small?

But the art of measurement would do away with the effect of appearance, and, showing the truth, would fain teach the soul at last to find rest in the truth, and would thus save our life. Would not mankind generally acknowledge that the art which accomplishes this result is the art of measurement?

Socrates (in *Protagoras*)

Part I
Growth

1
Abstracts of Previous Studies

NOTE: The following listings represent only a fraction of the literature on the measurements of horses, and are confined to papers giving actual measurements as well as descriptions. It will be noted that the great majority of the papers are by European (principally German) authors in reference to European breeds. However, the dimensional relationships existing in these breeds between height, weight, and girth are equally applicable to *all* breeds, including the light and heavy horses and ponies currently popular in the United States. In most of the papers, no mention is made as to whether the measurements of the adult male horses pertain to stallions or to geldings. In such cases the writer, for convenience, has referred to the animals simply as males (and the mares as females). While some of the information given in these abstracts may be of no direct use to the general breeder, it is included for the benefit of students of hippology who may find copies of the original papers difficult of access. From these studies the principal linear body measurements (along with weights, wherever they have been recorded) are given here in Tables 1 to 6, inclusive.

(Anon.) (1932). Measurements of height, chest girth, and front cannon (bone) girth in eleven English Derby candidates (3-year-olds). The average height is 63.85 inches (60.75-67.5), chest 70.4 inches (67.5-72.5), and cannon 7.91 inches (7.5-8.5). The list was in *Sporting Life* (issue not recorded).

(Anon.) (1937). Another list in *Sporting Life* similar to the above, of twenty-one 3-year-old candidates for the Epsom Derby of 1937. The average height is 64.29 inches (61-66.5), chest 69.81 (67-72), cannon 8.08 (7.75-9). The chest girths in

15

the foregoing two groups indicate an average body weight of about 1045 pounds.

Afanassieff, S. (1930). Numerous body measurements of male and female Orloff trotters, mostly at the ages of 3, 4, and 5 years. The combined number of horses measured at 3 and 4 years was 392, while at 5 years and over there were 261 stallions and 192 mares. See Tables 1 and 2.

Allen,—. (1905). This investigator, who is quoted by F. B. Morrison in the latter's book *Feeds and Feeding*, p. 926, found the average weight of Standardbred foals to be at one year 644 pounds; two years, 908 pounds; three years, 1026 pounds; and four years, 1102 pounds. These figures suggest that Allen's foals were somewhat overfed up to one year of age, then somewhat underfed.

Alminas, K. (1939). Detailed linear and angular measurements and weights of the ancient Lithuanian breed known as Zemaitukas, as found by Alminas and four other independent investigators of that breed. In addition, Alminas presents a review of the principal body measurements of twelve other breeds, ranging from the Huzulen (Polish) mountain pony to Belgian and English draft horses. There is also a bibliography of over 100 titles.

Bantoiu, C. (1922). Body measurements of 67 mature Orloff trotters.

Bilek, F. (1914). Body measurements of five desert-bred Arabs, thirteen purebred Arabs, and seven "halfbred" (Anglo-Arabs), with reference to their crossbreeding with Lipizzans.

Boicoianu, C. (1932). Heights (at withers) and trunk lengths of Belgian colts and fillies at one, two, three, and four years, along with more numerous measurements of adult horses. Also given are various measurements of the skull in three male and three female Belgians ranging from three to eight years of age. The average vertex (overall) skull length in stallions is 622 mm (24.50 in.) and in mares 618 mm

(24.33 in.). The frontal (maximum) skull width in stallions is 244.8 mm (9.64 in.) and in mares 242.9 mm (9.56 in.). Thus the heads of Belgian stallions and mares are almost identical in size.

Boing,— (1911). Twelve body measurements of 30 adult male and 32 adult female Belgian horses. The average heights are 63.82 inches in males and 62.72 inches in females; average chest girths 82.01 inches in males and 79.49 inches in females; and average cannon girths 10.20 inches in males and 9.06 inches in females. Weights, if estimated from chest girth, would average 1672 pounds in males and 1481 pounds in females. Thus these Belgian males average nearly a hundred pounds *more* than those recorded by von Nathusius (q.v.), while the females average at least 140 pounds *less*. Possibly von Nathusius's mares were in foal.

Brandes, H. (1926). Withers height, chest girth, chest depth, and cannon girth in 292 Trakehner mares, 48 Thoroughbred mares, 13 Thoroughbred stallions, 45 Trakehner stallions, and 197 halfbred stallions. The average height of the Trakehner stallions is 63.23 inches, chest girth 73.98, cannon 7.95, and weight 1264 pounds.

Brody, S. (1927). In this Research Bulletin 104 of the Missouri Agricultural Experiment Station, the author shows how the growth curves of many familiar animals (cattle, swine, sheep, rabbits, rats, mice, even chickens and pigeons) are essentially identical when proper allowances are made for the differing lengths of time that each species takes to reach full growth. The same type of growth curve is shown to apply also to the horse. The curve for humans is identical with that of the lower animals mentioned only from puberty onward, owing to the vastly longer period of prepubertal growth in man.

Butz, Henseler, and Schottler. (1921). Body measurements and weights of nine different breeds, as taken by Simon von Nathusius (q.v.) and quoted in their book, *Praktische Anleitung zum Messen von Pferden. Anleitungen der Deut-*

schen Gesellschaft fur Zuchtungskunde, 1921, Heft 2, p. 57. (See in Tables 1 and 2 the measurements attributed to von Nathusius.)

Caine,—. (1943). Weights of "liberally fed" Percheron and Belgian colts at birth (159 pounds), one year (1105 pounds), two years (1506 pounds), and three years (1619 pounds). At this rate of growth the expected weight at five years would be about 1740 pounds. This would be in typical ratio to the slightly high average birth weight of 159 pounds.

Crampton, E. W. (1923). Weights of draft colts from birth to four years. The total number of colts weighed was 409. In comparison with other groups, Crampton's colts, especially after the age of one year, were heavily overfed, since at only three years they weighed on the average 1790 pounds and at four years no less than 1980 pounds, although the average birth weight was only 120 pounds. A happy medium in growth rate and mature weight would be well below Crampton's 1980 pounds at four years and somewhat above that of Dawson, et al (q.v.), in which the average weight of geldings at five years was 1516 pounds.

Cunningham, K., and Fowler, S. H. (1961). Numerous body measurements and weights of Quarter Horse males and females from birth to five years. The number of colts measured at a particular age ranges from 3 to 13 (total, 66), and fillies from 6 to 60 (total, 188). While this is one of the most comprehensive studies of a specific breed made during recent years, certain of the measurements stated have had to be modified for my use, owing to differences in the respective measuring techniques employed. The trunk or body length, for example, in adult Quarter Horses is clearly more than a fraction of an inch greater than the height at withers, as recorded by these investigators; while the width of chest as stated by them is over 20 percent greater than that obtained by the usual manner of measuring.

Dawson, W. M. (1948). Body measurements of eight three-year-old male and ten three-year-old female Morgan horses. The data were supplied to me by Dawson from investigations carried on by the United States Department of Agriculture and co-operating stations. From these statistics on 3-year-olds it has been possible to estimate the probable respective measurements at five years, especially in view of the weights at the latter age being known from other sources. (See Linsley, D. C., and Dawson, et al, 1945).

Dawson, W. M., Phillips, R. W., and Speelman, S. R. (1945). Heights and weights at six-month intervals from weanling (six months) to five years of age, in male and female Belgian, Grade draft, Morgan, and light grade horses, raised under Western range conditions. The data refer to 297 horses which were foaled from 1927 to 1934 inclusive. Curves recording the increases in weight show seasonal ups and downs, due to losses sustained during winters when there was lessened pasturage. However, the general trends with age of both height and weight are fairly compatible with those found in other growth studies. It is notable that the draft horses grew apparently at the same rate as those of the light breeds.

Degen, K. (1933). Body measurements of 500 Oldenburg brood-mares. The Oldenburg is the heaviest of the German "warmblood" breeds.

Dimitriadis, J. N. (1937). Measurements of Skyros ponies, a breed indigenous to the Greek island of Skyros in the Sporades group. Body and head measurements are given of ten mares, and skull measurements of three mares and three stallions. The average withers height of the mares is 1053 mm (41.46 in.). The proportionate height of stallions, on the basis of comparative skull length, would be about 1064 mm (41.87 in.). No weights are listed.

Dinsmore, Wayne. (1939). Measurements of withers height and of chest height from ground in eight mature Thoroughbred stallions at Claiborne Stud (near Lexington, Kentucky) and six mature stallions at Idle Hour Stock Farm, 1935. At the former stud the average height was 65.375 inches, and average chest height from ground 35.50 inches (or 54.3 percent of the withers height). At the Idle Hour, the measurements were 63.833 and 34.375 inches, respectively (or 53.85 percent of the height). These figures show that leg length increases faster than withers height.

Dunn, N. K. (1973). Height, chest girth, and cannon girth in Arabian horses at the ranch of the California State Polytechnic University, Pomona. Selected as representative adult specimens were two stallions, fifteen geldings, and eleven mares. Mean heights and weights of the stallions are 60.2 inches and 987 pounds, geldings 60.02 inches and 998 pounds, and mares 58.07 inches and 963 pounds, respectively. The chest and cannon girths are correspondingly larger because of the heavier-than-usual bodyweights. Included also are the heights, weights, and girths of twenty newborn Arabs, the sexes being combined. The mean height is 37.10 inches, weight 85 pounds, chest 29.07 inches, and cannon 4.44 inches. These foals are therefore of typical size for newborn Arabs.

Ewart, J. C. (1901). Professor James Cossar Ewart, a Scottish animal geneticist, was a pioneer in the cross-breeding of domestic horses, zebras, and other equine species. In this paper he lists numerous body measurements of a foal (breed not stated) at monthly intervals from birth to one year, then at two years and three years, at which latter age a withers height of 57 inches was attained. While Ewart's table of measurements was first published in the *Live Stock Journal Almanac* in 1901, it has been repeated in several subsequent British books on horses, notably Axe's *The Horse, Its Treatment in Health and Disease* (1906), Vol. 8, p. 297. Possibly Ewart considered that little, if any, growth in height took place after the age of three years, whereas abundant data show that often a full inch is added between three and five years.

Feige, E. (1927). External body measurements of 5840 East Prussian mares measured between 1887 and 1902. The average withers height is 1654.3 mm (65.13 in.). The range in height is from 1530 to 1779 mm (60.24 in. to 70.04 in.). Thus there is a range of plus or minus 7.5 percent in relation to the average height. Unfortunately, height is the only measurement I have recorded on my list.

Flade, J. E. (1959). Numerous body and head measurements of Shetland stallions and mares, along with foals up to three years of age, bred in Germany. The number of stallions is 28, and mares 120. Possibly because of the relatively small number of stallions measured, the average height found in them is slightly less (1014 mm, or 39.92 in.) than in the mares (1022 mm, or 40.24 in.). The stallions, however, are larger in the girth of chest and, of course, the cannon. See Tables 1 and 2.

Franke, H. (1935). Various body measurements, along with lengths and inclinations of the fore and hind limb segments, in 186 Trakhener broodmares. The study was made to see whether there was any correlation between body measurements and length of stride. The correlation was found to be very slight in the case of bone lengths, and either very low or negative as regards limb angles. Presented also are a lesser number of measurements of 36 three-year-old Trakhener stallions and another series of 77 East Prussian stallions, which the author evidently differentiates.

Goreniuc, A. (1924). Body measurements, limb lengths, and limb inclinations in Noric (Pinzgauer, or Austrian draft) and Schleswig (Northwest German draft) stallions and mares. The numbers of each breed are not stated, and the measurements are of the two sexes combined. Both breeds are of about the same height (c. 1600 mm, or 63 in.), with the Schleswig being slightly the heavier in build.

Grange, E. A. A. (1894). A valuable, yet apparently little-known, series of measurements of four different breeds by an American investigator. The measurements were obtained of horses exhibited at the Columbian Exposition and the Michigan State Fair, both in 1893. The horses measured consisted of eleven Roadsters (Standardbreds), eight Coach horses, six "General Purpose" horses, and twenty-five draft horses. This is an especially useful study, since no fewer than 23 linear measurements, along with the weight, are recorded for each individual animal. Grange was one of the first, it would appear, to adapt and define a practical system of body measurements of the horse. His diagram, with minor modifications, is repeated here in Fig. 1.

Gregory, K. (1931). Body measurements and weights of 280 broodmares of the Hungarian saddle breed known as Furioso-North Star, or Mezohegyes. These fine riding horses are practically identical in size with the Hanoverian horse of Germany.

Gross, H. (1930). Average withers height, chest girth, cannon girth, and weight in 40 East Friesian stallions.

Gutsche,—. (1914). Numerous body measurements, limb segment lengths, and limb inclinations in two Trakehner colts and two Trakehner fillies at the ages of six months, one year, and four years. No weights are given.

Hering, A. (1925). Twelve standard body measurements of Rheinish-Dutch colts and fillies from birth to 2½ years, and in mature stallions and mares. No weights are given in either series. See Tables 3, 4, 5, and 6.

Hervey, J. (1941). Fourteen body measurements of six champion trotting stallions: Directum 4, Cresceus, The Harvester, Lee Axworthy, Spencer Scott, and Greyhound. The average height is 63.19 inches, average chest 70.96 inches, and average weight (as estimated from chest girth) 1092 pounds.

Hooper,—. (1921). In an article in *The Thoroughbred Record* for July 9, 1921, this author states that Thoroughbred fillies he observed at 11–13 months of age averaged 58 inches and 760 pounds, and colts of the same age 780 pounds. The weight of the fillies corresponds with a height of 57.5 inches; but both the height and weight imply an age of at least 13 months, which indicates that these foals were decidedly large for their age.

Iwersen, E. (1926). Twelve standard body measurements of Holstein colts and fillies from birth to five years. At birth the average withers height (sexes combined) is 1033 mm (40.67 in.) as compared with only 958.5 mm (37.74 in.) in the Rheinish-Dutch draft foals measured by Hering. See Tables 3, 4, 5, and 6.

Kolbe, W. (1928). Eight (averaged) standard measurements of four heavy breeds, all adult, as follows: 19 Norische (Pinzgauer) stallions and 24 mares, 6 Norman (French) stallions and 21 mares, one Cleveland Bay stallion and 16 mares, and 2 Clydesdale stallions and 5 mares. In addition, the same measurements are listed for 12 one-year-old Norische colts and 8 fillies, and 6 two-year-old Norische colts and 8 fillies. Kolbe also lists measurements of adult Pinzgauer stallions and mares as recorded by Dr. Wallner, and adult Clydesdale stallions and mares (along with mares only of the Rheinish-Belgian draft breed) as recorded by Simon von Nathusius (see Tables 1 and 2). The "Oberlander" horse mentioned by Kolbe is a lighter form of the Norische breed, native to Upper Bavaria at an elevation of about 5000 ft.

Kronacher, C., and Ogrizek, A. (1932). Inclinations of the various segments of the fore and hind limbs in 60 Brandenburg (semidraft) mares. Comparisons are made of these inclinations with those recorded by other authors on seven other breeds, ranging from Thoroughbreds to heavy drafters. The average withers height of the Brandenburg mares is 1638 mm (64.49 in.), ranging from 1570 mm to 1730 mm (61.81 in. to 68.11 in.). No weights are given. See also W. Kruger, following.

Kruger, W. (1939). Numerous linear body measurements and limb-segment lengths and inclinations of six adult male and female Trakehners and eight adult male and female Mecklenburg (heavy carriage) horses. This study, and that made by Kronacher and Ogrizek (q.v.), shows conclusively the fallacy of the "sloping" shoulders and pasterns so often specified in qualitative descriptions of desirable conformation in horses. For the inclination of the shoulder blade (scapula) from the horizontal Kruger gives 64 degrees for

Trakehners and 63 degrees for Mecklenburg horses. In 60 Brandenburg mares as measured by Kronacher and Ogrizek the average shoulder inclination is 58 degrees, ranging from 51 to 65 degrees. For the inclination (from the horizontal) of the front pasterns, Kruger gives 59 degrees for Trakehners and 67 degrees for Mecklenburgers. For the same part, Kronacher and Ogrizek find an average inclination of 55.5 degrees, with extremes of 44 and 66 degrees.

Kuffner, H. (1922). Numerous body measurements of Arab stallions and mares bred at Babolna, Hungary. Listed are the average measurements of three male and three female purebreds and five male and five female halfbreds (Anglo-Arabs). The average withers height of the purebred stallions is 1497 mm (58.94 in.), and of the purebred mares 1490 mm (58.66 in.). That of the halfbred stallions is 1540 mm (60.63 in.) and of the halfbred mares 1522 mm (59.92 in.). No weights are given.

Lesbre, F. X. (1893, 1894). Numerous body measurements and weights of several individual foals of both light and heavy breeds from birth to maturity. Given also are extensive measurements of a Belgian stallion as taken by Lavalard. A comparison is made of the various limb-bone lengths as averaged from five donkeys and two mules. A rare item of information is the body and limb measurements of a horse fetus of nine months, in which the height is 730 mm (28.74 in.), trunk length 600 mm (23.62 in.), and chest girth 590 mm (23.23 in.). The body weight for the latter chest girth would be 47 or 48 pounds.

Letard, E. (1925). Weights, and some heights, of foals of light and heavy breeds, at birth and at various ages up to 36 months. An especially valuable list is one of 18 Thoroughbred foals, averaged in height and weight at monthly intervals from birth to 16 months, during which period there is a gain in height of 940 mm (19.29 in.) and in weight of 330 kg (727.5 lbs).

Linsley, D. C. (1857). In this early volume the author presents much information on Morgan horses, including the heights and weights of numerous outstanding sires of the breed. From these listings I have averaged the heights and weights of 39 stallions picked at random. The mean height is 60 inches (55-65 in.) and the mean weight 1035 pounds (875–1,315 lbs). This combination of height and weight represents very well that of a horse of "average" size.

Madroff, C. (1935). In this and the three papers following, Madroff lists fifteen body measurements, which include withers height, croup height, trunk length, chest girth, chest width, chest depth, length and width of croup, length and width of head, girth of fore cannon, and bodyweight. Measurements—as in all European articles—are in the metric system. This first paper deals with Lipizzan horses—79 stallions and 128 broodmares—from breeding stations in Austria, Hungary, Czechoslovakia, Rumania, and Yugoslavia. These papers by Madroff are perhaps the best single sources of measurements of the breeds named, since they each comprise adequate numbers to furnish reliable average figures, and represent measurements that were taken by a systematic, uniform technique. See Tables 1 and 2.

———— (1936). Body measurements of Gidran (Anglo-Arabian) horses —89 stallions and 113 broodmares—from breeding stations in Hungary, Austria, Rumania, and Bulgaria. See note above regarding Lipizzan horses.

———— (1936). Body measurements of two breeds of the Nonius (Anglo-Norman) horse. Of the Large Nonius there are 20 stallions and 60 broodmares, and of the Small Nonius 68 stallions and 175 broodmares. See note above regarding Lipizzan horses.

———— (1937). Body measurements of two breeds of the Arab horse. Of purebred Arabs there are 51 stallions and 87 mares, and of halfbred (Anglo-Arabian) horses 46 stallions and 74 mares. These horses are from breeding stations in Yugoslavia, Rumania, Bulgaria, Hungary, and Czechoslovakia. See note above regarding Lipizzan horses.

Magiaru, C. (1936). Body measurements of six Lipizzan stallions and 45 Lipizzan mares. As will be noted from Tables 1 and 2, these Lipizzans are significantly smaller than those measured and weighed by C. Madroff in 1935.

Mieckley,—. (1894). 23 measurements of three Trakehner colts and three fillies at birth, one year, three years, and four–five years. The average withers height at birth in colts is 40.28 inches and in fillies 42.24 inches. The average weight at birth in colts is 107 pounds and in fillies 120 pounds. Evidently the small number of animals measured accounts for this sex discrepancy in size. In another group of 37 newborn foals (males and females combined) the average height is 41.37 inches and the average weight 113.3 pounds.

Müller, Dr. von. (1933). A study of the inclinations of the limb-segments in horses in relation to trotting and galloping speeds. Many measurements are listed, including withers height, chest girth, trunk length, etc. 237 horses in eight different categories are represented.

Nathusius, S. von. (1891, 1905, 1910, 1912). Numerous body measurements and percentage relationships (to withers height) in nine different breeds of horses. The numbers and breeds are as follows: 77 English Thoroughbred males and 33 females; 421 East Prussian males and 31 females; 497 Hanoverian males and 98 females; 12 Holstein males and 27 females; 124 Oldenburg males and 48 females; 35 East Friesian males; 61 Danish North-Schleswig females; 112 purebred Belgian males and 76 females; 48 "English draft" (Clydesdale or Shire?) males and 28 females. With such a preponderance of "males," these horses are probably geldings. These series of measurements by von Nathusius are quoted also in a number of German textbooks on animal husbandry. They are perhaps the most comprehensive and reliable lists ever recorded of the breeds in question.

Nicolescu, J. (1923). Body measurements and limb angulations in 95 male and 105 female Hanoverian horses. The withers heights average the same (1626.5 mm, or 64 in.) as in the Hanoverians measured by von Nathusius (see Tables 1 and 2), but show a marked sex difference: 1642 mm or 64.65 inches for males, and only 63.42 inches for females.

Ogrizek, A. (1923). Eighteen body measurements of seven male Veglia ponies ranging in age from eight years to over thirty years, and one female aged twenty-one years. Like most other pony breeds, evidently the Veglia is a hardy, long-lived animal.

Osowicki, A. (1904). Height, chest girth, and cannon girth in male and female Huzulen ponies (the numbers of each sex not stated). In these measurements the males average 52.60, 61.97, and 6.86 inches, respectively; and the females 51.65, 63.42, and 6.43 inches. Weights, if estimated from chest girth, would average 728 pounds in males and 748 pounds in females. The Huzulen is a Polish mountain pony.

Plischke, A. (1927). 65 body measurements and nine limb angulations in five male and five female Thoroughbreds at 1, 2, 3, 4, and 5 years, respectively. For comparison are given the same measurements in two male and two female Trakehner (East Prussian) horses at ½ year and one year, and two males and four females at 4 years and older.

Radescu, T. (1923). Sixteen body measurements and six limb angulations in 90 (combined) male and female Thoroughbreds.

Rhoad, A. O. (1928). Seven body measurements on a total of 1395 draft horses and mules tested for pulling power. The 1395 animals consist of 759 Percherons, 256 Belgians, 55 Clydesdales, 48 Shires, 38 mules, and 239 others. As to sex, there are six stallions, 450 mares, 867 geldings, 38 mules (sex not stated), and 34 unknown. The average height is 65.48 inches, chest girth 79.87 inches, fore cannon 9.66 inches, hind cannon 10.83 inches, and weight 1523 pounds. The heights range from 55 to 74 inches, and the weights from 1000 to 2050 pounds. Various coefficients of correlation are given.

Rösiö, B. (1928). Withers height, chest girth, and cannon girth in male and female North Swedish horses. The numbers of each sex are not given. The foregoing measurements in males are 60.63, 75.59, and 8.66 inches, respectively; and in females 59.84, 73.23, and 8.27 inches. Weights, if estimated from chest girth, would average 1321 pounds in males and 1158 pounds in females.

Schilke, F. (1922). Six body measurements, including height, chest girth, and cannon girth, in Trakehner (East Prussian) horses. In males the foregoing measurements average 63.66, 75.83, and 8.19 inches, respectively; and in females 62.91, 74.02, and 7.52 inches. The numbers of each sex are not given. Weights, if estimated from chest girth, would average 1322 pounds in males and 1196 pounds in females. These measurements should be compared with those recorded by von Nathusius (1891) for Trakehners listed here in Tables 1 and 2.

Schöttler, F. (1910). Withers and croup heights, chest girth, trunk length, and cannon girth in male and female Hanoverian horses. The average withers height in males is 64.67 inches, croup height 64.58 inches, chest girth 76.69 inches, trunk length 65.24 inches, and cannon girth 8.35 inches. The corresponding measurements in females are 63.43, 63.21, 76.70, 64.78, and 7.83 inches, respectively. Weights, if estimated from chest girth, would average 1368 pounds in males and 1331 pounds in females. These estimated weights are considerably higher than those recorded for Hanoverian horses by von Nathusius (see Tables 1 and 2).

Solanet, E. (1946). This book, by a noted hippologist of Argentina, includes height, chest, and cannon measurements of Criollo (native) horses, but no body weights. The average height in males is 56.85 inches, chest 68.90 inches, and cannon 7.56 inches, respectively. Each of these measurements is approximately 95 percent as large as the corresponding ones in Quarter Horses, showing that the body build of the Criollo has resulted from performing the same kind of work (e.g., as a cow pony) as the Quarter Horse. In Criollo mares the average height is 56.26 inches, chest 69.41, and cannon 7.16. Weights, if estimated from chest girth, would average 992 pounds in males and 986 pounds in females.

Stang, V., and Wirth, D. (1926). Tables from other authors giving the body measurements of various breeds of horses. In: *Tierheilkunde und Tierzucht,* Vol. 1. Berlin.

Stegen, H. (1929). Thirteen body measurements on a total of 364 male and female Hanoverian horses from birth to "full grown" (5 years plus). These measurements, for various ages, are listed here in Tables 3, 4, and 5 (the sexes being combined). Estimated bodyweights are in Table 6.

Stratul, J. (1922). Sixteen body measurements and six limb angulations in 102 (combined male and female) Thoroughbreds. Measurements are given also for Stratul's "10 best" Thoroughbred stallions. The principal measurements of the latter are: withers height 64.00 inches, chest girth 70.16 inches, and cannon girth 7.75 inches. The weight, as estimated from chest girth, would be only 1047 pounds, which suggests that these Thoroughbreds were 3-year-olds.

Svanberg, V. (1928). Thirteen body measurements of Finnish horses —314 stallions and 779 mares. The average measurements of the stallions are: withers height 60.67 inches, chest girth 71.53 inches, and cannon girth 7.99 inches. In the same measurements the mares are 59.96, 70.90, and 7.60 inches, respectively. Weights, if estimated from chest girth, would average 1110 pounds in stallions and 1052 pounds in mares. Thus these Finnish horses are approximately the same size as the Lipizzaners measured by C. Madroff (see Tables 1 and 2). According to Svanberg, the Finnish horse, during the last 200 years, has increased about 5½ inches in withers height and become correspondingly larger all over.

Szabo,—. (1924). Sixteen body measurements and eight limb angulations on Hungarian mares of the Small Nonius breed (number not given). The average withers height is 61.50 inches, chest girth 71.31 inches, and cannon 7.63 inches. Weight, if estimated from chest girth, would average 1070 pounds. These Small Nonius mares are thus considerably smaller than those measured by C. Madroff and listed here in Table 2.

Team, C. B. (1949). Height, weight, chest girth, and cannon girth in 25 Arab stallions and 25 mares at the Kellogg Arabian Horse Ranch in Pomona, California. See Tables 1 and 2. Major Team also supplied me with the foregoing measurements in six newborn Arab colts and five newborn fillies.

In the colts the average height was 38.6 inches, chest girth 30.75 inches, cannon girth 4.40 inches, and weight 94 pounds. In the fillies these measurements were 38.2 inches, 30.25 inches, 4.35 inches, and 93 pounds, respectively.

Trowbridge, E. A., and Chittenden, D. W. (1930). Weights, at daily intervals during the first nine to twelve months, of three Percheron colts and three fillies.

——— (1932). The same foals as measured in 1930, plus one additional foal (sex?), measured at birth, weaning time, and one, 2, 3, and 5 years, respectively. 27 body measurements at each age are given. See Tables 3, 4, 5, and 6.

Ullmann, G. (1939). Measurements of weight, height, neck and trunk length, chest girth, and cannon girth, from birth to 48 days, in two very small Shetland ponies, a male and a female, in the Zoological Garden at Cologne, Germany. Evidently the foals were undersized at birth, since at the end of 30 days their weights had each more than doubled.

Unckrich,—. (1925). Nine body measurements in mares of the German breed known as Zweibrücker (number not stated). The average height is 63.27 inches, chest girth 75.39 inches, and cannon girth 8.07 inches. Weight, if estimated from chest girth, would average 1264 pounds. This would make the Zweibrücker mares about the same size and build as female Hanoverians.

Voltz, W. (1913). Thirty body measurements of seven East Prussian foals (sexes combined) at six months, 1½ years, and 2½ years. At these respective ages the heights are 49.09, 53.98, and 57.80 inches; chest girths 52.64, 59.17, and 66.58 inches; and cannon girths 6.14, 6.89, and 7.28 inches. Weights, if estimated from chest girth (the sexes being combined) would average 449 pounds at six months, 628 pounds at 1½ years, and 882 pounds at 2½ years, respectively. These figures indicate a very slow rate of growth (cf. Table 6).

Weiss-Tessbach, A. (1923). Fifteen body measurements of Pinzgauer (Austrian draft) horses at one year (32 colts, 15 fillies), two years (16 colts, 8 fillies), three years (one colt, 9 fillies), and adult (10 stallions, 2 geldings, 42 mares). The adult withers height is 64.88 inches for males and 64.17 inches for females; chest girth 77.95 inches for males, 76.97 inches for females. The chest girths indicate body weights of about 1436 pounds for males and 1338 pounds for females. Thus these particular Pinzgauers are very light in weight compared with their quite typical draft-horse heights.

Wiechert, F. (1927). Withers height and nine limb-bone and pelvic length measurements in 46 geldings and 20 mares, distributed as follows: 17 distance runners, 10 hunters, 22 jumpers, and 17 dressage (parade) horses. The average heights and chest girths of the geldings are 63.98 and 71.63 inches, respectively; and of the mares 62.77 and 70.33 inches, respectively. The actual average body weights are 1041.5 pounds in the geldings and 977.7 pounds in the mares. These weights suggest that the horses are 3-year-olds.

Wilkomm, W. (1921). Ten body measurements of the German breed known as Beberbecker (mares only; number not stated). The average withers height is 63.11 inches, chest girth 73.62 inches, and cannon girth 7.91 inches. The actual weight (average) is 1196 pounds; as estimated from chest girth, 1176 pounds.

Willoughby, D. P. (1950). Thirty-five body measurements taken by me on each of 16 Arabian horses: 3 stallions, 9 broodmares, 2 colts, and 2 fillies, on the Arabian horse ranch (El Cortijo) of Donald and Charles McKenna, Claremont, California. The average height of the stallions is 58.74 inches, chest girth 68.76 inches, and weight (as estimated from chest girth) 986 pounds. The corresponding measurements of the mares are 58.45 inches, 73.02 inches, and 1148 pounds, respectively. The manifest "overweight" status of the mares is due to nearly all of them being in foal (approximately 300 days). The stallions also are heavier-built than is typical of the breed. With these conditions being taken into account, these detailed body measurements of both young and mature Arabian horses, along with those of C. Madroff, et al (q.v.), have proven invaluable to the writer in arriving at optimum standards of external conformation in this breed.

Wöhler, H. (1927). Nine body measurements of 200 Hanover broodmares. See Table 2.

Zimmermann, C. (1933). Twelve body measurements of numerous Rheinish-Belgian draft colts and fillies from birth to adult. For some of the principal measurements and growth ratios (sexes combined), see Tables 3, 4, 5, and 6.

2
The Physique of the Horse in Various of Its Breeds

THE AVERAGE BODY MEASUREMEMENTS OF NUMEROUS BREEDS OF ADULT horses are listed in Tables 1 and 2.* The measurements are here, initially, confined to withers (shoulder) height, chest girth, cannon (of the front leg) girth, and body weight. As in many of the listings large numbers of horses (sometimes several hundred) are represented, the resulting average measurements may be accepted as reliable, especially so far as height, weight, and cannon girth are concerned. Chest girth is another matter, and shows marked variability among the figures recorded by different investigators. This variability may be due to several factors: faulty *placement* of the tape (it should always be *behind* the withers, as indicated in Fig. 1) ; too much or too little pull on the tape; or the use of a fabric tape that is either *shrunken* (which would give too large a chest girth) , or is *stretched* (which would give too small a girth) . Despite the latter difficulties, it has been found that so long as the chest girth is taken with a *consistent* technique, it provides a remarkably reliable measurement for the estimation of body weight. If the practical breeder will compare his measurings with the chest girths and body weights listed in Table 7, he should quickly learn whether or not he is measuring in a consistent manner. Also, such breeders as are not interested in statistics may proceed directly to Table 8. However, Tables 1 to 6 inclusive are full of interesting information, some of which may now be examined.

The emphasis here placed on chest girth is because this readily applicable measurement can be used to estimate *body weight,* which usually is unknown and/or cannot be determined because of the absence of a suitable weighing scales. And weight, of course, is the best

---

* Since this book is directed primarily to horsemen in the United States, all measurements are given in the English system—that is, in inches and pounds.

33

TABLE 1.
Average Heights, Girths, and Weights in LIGHT ("Warmblood") Horses.
(Males)   ( ) Est. by writer.

Breed	No. of Examples	Source	Withers Height, in.	Chest Girth, in.	Cannon Girth, in.	Bodyweight, lbs. Observed	Bodyweight, lbs. Estimated*	Difference, percent, Observed from Estimated Weight
Arabian	51	Madroff, 1937	59.69	67.51	7.35	901	933	−3.43
Arabian	25	Team, 1949	59.27	68.20	7.41	922	962	−4.16
Anglo-Arab	46	Madroff, 1937	60.02	69.10	7.44	967	1000	−3.30
English Thoroughbred	77	von Nathusius, 1891	63.11	71.77	7.99	1131	1121	+0.86
Standardbred	10	Grange, 1894	62.80	70.24	7.99	1130	1051	+7.52
Orloff Trotter	261	Afanassieff, 1930	62.05	68.07	7.64	958	957	+0.10
Quarter Horse	13	Cunningham & Fowler, 1961	60.00	73.00	7.90	1201	1180	+1.80
Lipizzan	6	Magiaru, 1936	58.66	66.86	7.40	937	907	+3.35
Lipizzan	79	Madroff, 1935	60.08	71.26	8.03	1102	1097	+0.46
Gidran	89	Madroff, 1936	62.09	71.77	8.03	1062	1121	−5.30
Large Nonius	20	Madroff, 1936	65.20	78.35	8.62	1360	1459	−6.77
Small Nonius	68	Madroff, 1936	62.44	74.88	8.23	1144	1273	−10.10
East Prussian	421	von Nathusius, 1891	63.54	73.70	8.27	1222	1214	+0.68
East Prussian	46	Wiechert, 1927	63.98	71.63	(7.99)	1042	1114	−6.55
Zemaitukai	7	Alminas, 1937	56.30	66.54	7.68	902	894	+0.89
Coach	7	Grange, 1894	63.74	73.86	8.43	1277	1222	+4.50
"All Purpose"	6	Grange, 1894	65.51	75.83	8.66	1393	1323	+5.31
Hanoverian	497	von Nathusius, 1912	64.06	74.76	8.54	1296	1267	+2.29
Holstein	12	von Nathusius, 1891	64.09	75.12	8.58	1305	1285	+1.55
Oldenburg	124	von Nathusius, 1891	64.45	76.97	9.02	1398	1383	+1.12
Av., 10 Lighter Examples		———	60.494	69.118	7.696	990	1000	−1.04
Av., 10 Heavier Examples		———	63.614	74.824	8.423	1270.5	1269	+0.12

Draft Breeds

Breed	No. of Examples	Source	Withers Height, in.	Chest Girth, in.	Cannon Girth, in.	Bodyweight, lbs. Observed	Bodyweight, lbs. Estimated*	Difference, percent, Observed from Estimated Weight
East Friesian	35	von Nathusius, 1891	64.69	80.47	9.41	1490	1580	−5.70
Percheron	1138	Rhoad, 1928	65.48	79.87	9.66	1523	1545	−1.42
Percheron	7	Trowbridge and Chittenden, 1932	65.00	79.50	9.58	1500	1524	−1.57
Belgian	112	von Nathusius, 1891	63.70	80.47	9.80	1576	1580	−0.25
Clydesdale	48	von Nathusius, 1891	64.41	80.12	10.47	1702	1560	+9.10
"Draft"	10	Grange, 1894	65.67	85.43	10.47	1885	1891	−0.32
Average (6)		———	64.825	80.98	9.90	1609	1610	−0.09

Shetland Pony

Breed	No. of Examples	Source	Withers Height, in.	Chest Girth, in.	Cannon Girth, in.	Bodyweight, lbs. Observed	Bodyweight, lbs. Estimated*	Difference, percent, Observed from Estimated Weight
Shetland	28	Flade, 1959	39.92	51.06	5.51	389	404	−3.66

*From the author's formula: Bodyweight, lbs. (adult males) = (.14475 Chest Girth, in.) [3]

TABLE 2.
Average Heights, Girths, and Weights in LIGHT ("Warmblood") Horses.
(Females)

Breed etc.	No. of Examples	Source	Withers Height, in.	Chest Girth, in.	Cannon Girth, in.	Bodyweight, lbs.		Difference, percent, Observed from Estimated Weight
						Observed	Estimated*	
Arabian	87	Madroff, 1937	58.96	66.93	7.12	952	884	+7.80
Arabian	25	Team, 1949	58.99	69.20	7.04	922	977	−5.56
Anglo-Arab	74	Madroff, 1937	59.29	68.46	7.13	996	946	+5.42
Standardbred	1	Grange, 1894	62.00	68.50	7.50	1000	948	+5.60
Orloff Trotter	192	Afanasieff, 1930	61.77	68.94	7.48	958	966	−0.71
Quarter Horse	60	Cunningham & Fowler, 1961	58.50	73.10	7.40	1172	1152	+1.85
Lipizzan	45	Magairu, 1936	58.34	66.81	7.26	900	879	+2.46
Lipizzan	128	Madroff, 1935	59.17	70.16	7.64	1015	1018	−0.23
Gidran	113	Madroff, 1936	61.50	71.40	7.72	1017	1073	−5.16
Large Nonius	60	Madroff, 1936	65.00	76.86	8.24	1372	1339	+2.58
Small Nonius	175	Madroff, 1936	62.32	73.96	7.87	1162	1193	−2.49
East Prussian	31	von Nathusius, 1891	63.46	76.06	7.83	1195	1297	−7.81
East Prussian	20	Wiechert, 1927	62.60	71.06	7.76	1011	1058	−4.36
Coach	1	Grange, 1894	64.75	74.50	8.00	1250	1219	+2.63
Beberbeck	?	Wilkomm, 1921	63.11	73.62	7.91	1196	1176	+1.78
Hanoverian	98	von Nathusius, 1912	64.02	77.16	8.03	1288	1354	−4.83
Hanoverian	200	Wöhler, 1927	62.91	75.51	8.03	1311	1269	+3.38
Zemaitukai	26	Alminas, 1937	55.12	65.75	6.93	910	838	+8.69
Holstein	27	von Nathusius, 1891	64.65	76.93	8.07	1329	1300	+2.23
Oldenburg	48	von Nathusius, 1891	64.61	79.29	8.50	1400	1469	−4.66
Av., 10 Lighter Examples		————	59.774	68.721	7.357	967	957	+1.04
Av., 10 Heavier Examples		————	63.333	75.699	7.988	1266	1275	−0.70

Draft Breeds

Clydesdale	28	von Nathusius, 1891	64.37	79.92	9.68	1561	1506	+3.67
Belgian	76	von Nathusius, 1891	63.78	81.97	9.21	1634	1623	+0.67
Belgian	?	Boicoianu, 1932	64.69	82.91	9.72	1653	1681	−1.65
"Draft"	5	Grange, 1894	65.10	83.00	9.25	1703	1687	+0.97
Average (4)		————	64.485	81.95	9.47	1637	1623	+0.86

Shetland Pony

Shetland	120	Flade, 1959	40.24	50.51	5.28	381	380	+0.26

*From the author's formula: Bodyweight, lbs. (adult females)=(.14341 Chest Girth, in.)^3

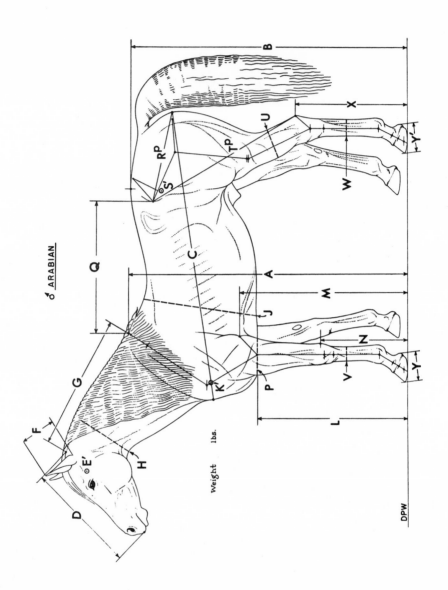

Fig. 1. Standard body measurements of the horse, as referred to in this study. The figure shown is that of an Arabian stallion (for actual dimensions, see Table 9.) The measurements shown are as follows:

A - *Height of withers*
B - *Height of croup*
C - *Length of trunk, slantwise, from shoulder to buttock*
D - *Length of head (total)*
E' - *Width of forehead, maximum*
F - *Length of ear (externally)*
G - *Length of neck, from poll to point of withers*
Ⓗ - *Girth of neck, minimum, at throatlatch*
Ⓙ - *Girth of chest, maximum, behind withers*
K' - *Width of shoulders, maximum, across heads of humeri*
L - *Height of chest from ground*
M - *Height of elbow (top of olecranon)*
N - *Height of knee (top of pisiform)*
Ⓟ - *Girth of forearm, maximum*
Q - *Length of back, from point of withers to a line connecting the most anterior points of the two ilia*
R^P - *Projected* slantwise length of croup, from point of hip to point of buttock*
S' - *Width of croup, maximum, across heads of femora*
T^P - *Projected* length from point of hip to point of hock*
Ⓤ - *Girth of gaskin, maximum*
Ⓥ - *Girth of fore cannon, minimum*
Ⓦ - *Girth of hind cannon, minimum*
X - *Height of hock (top of calcaneum)*
Y - *Length of fore hoof (sole)*
Y' - *Length of hind hoof (sole)*
Z - *Width of fore hoof* ⎫
Z' - *Width of hind hoof* ⎬   *not shown on drawing*

**Projected here means brought into line with the midsagittal plane of the body, as seen in a straight profile view.*

single measure or index of physical growth (as well as basis for estimating the feed needs of the growing foal, or foals).

My formulas for predicting body weight from chest girth in adult male and female horses (irrespective of breed) have been derived from the observed or actual weights listed and averaged in Tables 1 and 2. The resulting correspondence between estimated and observed weights is, on the average, remarkably close. This is at once seen from a comparison of the two sets of figures, especially when it is considered that adult horses of both sexes and a wide range of sizes —from Shetlands to draft breeds—are accommodated by the formulas. These formulas, for *adult* horses—or those having chest girths of 66 inches or larger—are as follows:

(Males) Body weight, lbs. = $(.14475 \text{ Chest Girth, inches})^3$

(Females) Body weight, lbs. = $(.14341 \text{ Chest Girth, inches})^3$

Since it has not been possible to separate the weights of stallions and geldings, the single formula for "males" has herein been applied to both groups.

The body weights of immature horses—that is, colts and fillies from birth to the age of five years, or having chest girths of less than 66 inches—are herein estimated from the formulas:

(Colts)

Body weight, lbs. = $(.1387 \text{ Chest Girth, inches, } + 0.400)^3$

(Fillies)

Body weight, lbs. = $(.1382 \text{ Chest Girth, inches, } + 0.344)^3$

It should be noted that while the weights of immature Shetland ponies are estimated by the latter formulas, the weights of Shetlands having chest girths of 50 inches or larger are estimated by the formulas given previously for *adult* horses. The necessity of having separate formulas for immature and adult horses is because the adults, for a given girth of chest, weigh slightly *less* than the foals.

In all the four foregoing formulas, the symbol [3] at the end—which indicates that the derived measurement within the parentheses is to be *cubed*—is necessary in order to convert the linear measurement of chest girth into the *cubic* or volumetric measurement of weight.*

---

* It should be pointed out that certain other measurements (e.g., cannon girth), or *combinations* of measurements (e.g., chest girth *and* cannon girth), give in many instances even better predictions of body weight than does chest girth alone. But since the purpose of this book is to provide a simple, practical method for use by the general breeder, the use of more involved (and complex) formulas, along with the endless presentation of graphs (which few readers know how to interpret) has purposely been avoided. For the information of such readers as wish to know the formulas used by the writer in establishing typical size-relationships between the principal measurements of the body, details are given in the Appendix.

It is certainly remarkable, in the weight-prediction formulas given above, that such a wide range in weight can be accommodated simply by the use of chest girth alone. One might suppose, for instance, that in heavy (draft) breeds a different formula would have to be used than in light breeds such as Arabs and Thoroughbreds, yet evidently the long trunk of the draft horse is counterbalanced by his relatively short legs, so that on the basis of his chest girth he weighs the same as a horse of any of the light (and long-legged) breeds. Similarly, the small chest and short trunk of the young foal are compensated by his relatively long and heavy legs.

We may now briefly consider Tables 3, 4, 5, and 6, respectively, in that order.

Growth in withers height is measured with a calibrated wooden standard, especially designed for thus measuring horses, ponies, dairy cattle, and other farm animals. The height is measured perpendicularly from the ground to the highest point on the withers. In order to insure perpendicularity, some measuring standards are equipped with a spirit level.*

Hippologists concerned with the proportions or conformation of the body commonly use withers height as a basic measure to which to relate other body and limb measurements, as for example trunk length, head length, forelimb length (elbow height), croup and shoulder widths, etc. However, as a measure of general body nutrition, withers height is of little or no assistance, since the height of undernourished animals may equal, or even exceed, that of well-fed ones. For this reason, a much more reliable index of nutrition in horses (or cattle) is body weight. The latter, where facilities permit, may either be measured direct, or, as an alternative, estimated from chest girth by means of the formulas presented above. And in order to insure that an animal is not gaining in weight simply by putting on excess fat, it is well to use the cannon girth as a check on what the proportionate chest girth (and body weight) *should be.* The interrelationships of the latter three measures are listed in Table 7.

In Table 3 it will be noted that horses of both sexes and both light, medium, and heavy breeds are represented. The *rate* of growth in height, as indicated by the figures in Table 3, shows no consistent difference between light and heavy breeds of horses. Neither, when the figures for colts and for fillies are separated, is any typical sex-

---

* One manufacturer of such a measuring standard is the Miller Harness Company, Inc., 131 Varick Street, New York, N.Y. 10013.

TABLE 3

Height at Withers (inches) in Foals in Relation to Mature Height (100%) *

( ) Estimated

Breed and Author	Birth	1 mo.	3 mos.	6 mos.	1 year	1½ yrs.	2 yrs.	3 yrs.	4 yrs.	5 yrs.
Rheinish-Dutch (Hering, 1925)	37.74 *59.31*	42.38 *66.62*	47.44 *74.57*	51.79 *81.40*	56.44 *88.71*	58.27 *91.58*	60.59 *95.24*	62.46 *98.17*	(63.17) *(99.29)*	63.62 *100.00*
Rheinish-Belgian (Zimmermann, 1933)	39.35 *62.72*	43.03 *68.59*	48.15 *76.74*	52.01 *82.90*	56.58 *90.18*	58.50 *93.25*	60.63 *96.64*	61.71 *98.36*	(62.34) *(99.36)*	62.74 *100.00*
Percheron Colts, Av. of both groups (Trowbridge and Chittenden, 1932)	41.24 *63.45*	(45.04) *(69.29)*	(49.93) *(76.82)*	(54.15) *(83.31)*	58.42 *89.88*	(60.95) *(93.77)*	62.50 *96.15*	64.50 *99.23*	(64.81) *(99.71)*	65.00 *100.00*
Belgian (Boicoianu, 1932)	(41.56) *(64.23)*	45.45 *70.25*	50.83 *78.56*	55.33 *85.52*	60.02 *92.76*	(61.76) *(95.47)*	62.83 *97.11*	63.84 *98.66*	64.21 *99.24*	64.70 *100.00*
Light and Draft combined (Dawson et al, 1945)	(38.88) *(62.80)*	(42.52) *(68.67)*	(47.19) *(76.21)*	50.80 *82.04*	55.05 *88.92*	57.41 *92.72*	58.64 *94.70*	60.37 *97.50*	61.38 *99.13*	61.92 *100.00*
Hanoverian (Stegen, 1929)	40.39 *62.46*	43.86 *67.82*	48.50 *75.01*	51.97 *80.37*	56.16 *86.85*	59.72 *92.36*	61.22 *94.73*	63.27 *97.84*	64.21 *99.30*	64.67 *100.00*
Holstein (Iwersen, 1926)	40.67 *63.71*	(43.97) *(68.88)*	48.21 *75.52*	51.83 *81.19*	55.34 *86.68*	58.88 *92.23*	61.22 *96.40*	62.54 *97.96*	63.45 *99.35*	63.84 *100.00*
Quarter Horse (Cunningham and Fowler, 1961)	35.80 *60.42*	(39.50) *(66.67)*	44.25 *74.68*	49.40 *83.38*	53.65 *90.55*	56.50 *95.36*	57.50 *97.05*	58.10 *98.06*	59.00 *99.58*	59.25 *100.00*
Av., above 8 series	39.45 *62.40*	43.22 *68.36*	48.06 *76.03*	52.16 *82.51*	56.46 *89.31*	59.00 *93.33*	60.64 *95.92*	62.10 *98.24*	62.82 *99.38*	63.22 *100.00*
Ratios adopted as Optimum	62.50	68.46	76.11	82.72	89.40	93.41	95.87	98.62	99.35	100.00

*The upper line in each breed gives the height in inches (usually as converted from millimeters). The lower line gives the *percentage* relationship of the height to that at 5 years of age.

difference in the growth rate shown. For these reasons it is assumed herein that the typical, or normal, rate of growth in height is the same, on the average, for both sexes and all breeds. Departures from this typical rate, which for various reasons are only to be expected, may therefore be regarded as accidental deviations from the theoretical normal trend of growth. While height variations as great as ± 6 percent from the average for a given age may occur, usually the degree of variation is not over ± 3 percent. The Quarter Horse at birth, for example, shows a height about 4 percent *less* than would be expected from its mature height.

Actually, the growth ratios shown in the bottom row in Table 3 should not be taken as more than a general trend, since the ratio of height at birth to height at maturity varies in accordance with the *size* of the breed. On the average, the expected height at birth is equal to 0.568 × the height of the parent (of the same sex) + 3.60 inches. This formula yields the following heights for newborn males: Shetland, 26.10 inches; Arab, 37.50 inches; Thoroughbred, 39.83 inches; draft, 40.43 inches. The corresponding birth/adult height ratios are: Shetland 65.90, Arab 62.81, Thoroughbred 62.45, and draft 62.35. However, such breed deviations in the ratio of withers height need not be involved in the practical estimation of nutritional needs proportionate to general body size.

In Table 4 is shown the growth in *cannon girth* ("bone") among the same six breeds of horses as are listed in Table 3. The cannon girth—which may be taken as a general index of the caliber or *thickness* of the bones of the limbs—grows at a rate intermediate between that of the withers height (which grows fastest) and the chest girth, to which the cannon girth is closely related proportionwise (see Fig. 2). For various formulas that express the ratio of cannon girth to chest girth in light horses, draft horses, and Shetland ponies, respectively, see the Appendix. In Table 4, the "Ratios adopted as optimum" are only a general average of both sexes and various breeds, and cannot be applied to a specific sex and breed (light or heavy). For the latter purpose, use Table 7. At all ages and in both light, draft, and pony breeds, cannon girth is larger in males than in females, progressively more so as age and general body size increase.

Table 5 shows the increase in chest girth with age in the same six breeds as are listed in Tables 3 and 4. In this measurement the trend of growth in the two sexes combined may safely be taken as the trend in each sex separately, since the average girth at a given age is almost identical in colts and fillies. Indeed, in mature stallions

TABLE 4

Girth of Cannon (inches) in Relation to Mature Cannon Girth (100%)

Sexes combined

( ) Estimated

Breed and Author	Birth	1 mo.	3 mos.	6 mos.	Age 1 year	1½ yrs.	2 yrs.	3 yrs.	4 yrs.	5 yrs.
Rheinish-Dutch (Hering 1925)	5.53	6.02	6.73	7.40	8.15	8.80	9.04	(9.28)	(9.51)	9.67
	57.22	*62.42*	*69.66*	*76.56*	*84.32*	*91.05*	*93.48*	*(96.02)*	*(98.40)*	*100.00*
Rheinish-Belgian (Zimmermann 1933)	6.04	6.46	7.23	7.89	8.94	9.35	9.80	10.08	(10.38)	10.59
	57.06	*60.97*	*68.31*	*74.54*	*84.39*	*88.30*	*92.57*	*95.18*	*(98.02)*	*100.00*
Percheron Colts, Av. of both groups (Trowbridge and Chittenden, 1932)	6.00	(6.46)	(7.20)	(7.87)	8.63	(9.07)	9.35	9.42	(9.51)	9.58
	62.63	*(67.43)*	*(75.16)*	*(82.15)*	*90.08*	*(94.68)*	*97.60*	*98.33*	*(99.27)*	*100.00*
Hanoverian (Stegen, 1929)	4.96	5.41	6.12	6.46	7.03	7.76	7.85	8.23	8.36	8.46
	58.60	*63.95*	*72.19*	*76.28*	*83.02*	*91.63*	*92.79*	*97.21*	*98.77*	*100.00*
Holstein (Iwersen, 1926)	5.10	(5.59)	6.32	6.81	7.19	7.95	8.33	8.46	8.59	8.72
	58.46	*(64.11)*	*72.46*	*78.10*	*82.39*	*91.19*	*95.49*	*97.02*	*98.50*	*100.00*
Quarter Horse (Cunningham and Fowler, 1961)	4.45	(4.99)	5.55	6.25	6.90	7.35	7.35	7.35	7.45	7.65
	58.17	*(65.23)*	*72.55*	*81.70*	*90.20*	*96.08*	*96.08*	*96.08*	*97.39*	*100.00*
Av., above 6 series	5.35	5.84	6.52	7.11	7.81	8.38	8.62	8.80	8.97	9.11
	58.30	*64.10*	*71.62*	*78.08*	*85.68*	*91.99*	*94.62*	*96.62*	*98.43*	*100.00*
Ratios adopted as Optimum	*58.30*	*64.63*	*72.22*	*78.15*	*85.51*	*90.50*	*94.00*	*97.31*	*99.34*	*100.00*

TABLE 5
Girth of Chest (inches) in Relation to Mature Chest Girth (100%)
Sexes combined
( ) Estimated

Breed and Author	Birth	1 mo.	3 mos.	6 mos.	1 year	1½ yrs.	2 yrs.	3 yrs.	4 yrs.	5 yrs.
Rheinish-Dutch (Hering, 1925)	34.47	41.99	50.08	57.68	66.12	72.68	76.61	(78.57)	(80.94)	81.60
	42.24	*51.46*	*61.37*	*70.69*	*81.03*	*89.07*	*93.88*	*(96.90)*	*(99.19)*	*100.00*
Rheinish-Belgian (Zimmermann, 1933)	37.46	43.90	51.75	59.11	69.07	73.74	79.11	83.03	(83.89)	84.19
	44.49	*52.14*	*61.47*	*70.21*	*82.04*	*87.59*	*93.97*	*98.62*	*(99.64)*	*100.00*
Percheron Colts, Av. of both groups (Trowbridge and Chittenden, 1932)	35.45	(42.50)	(50.76)	(57.29)	65.41	(71.55)	74.77	78.19	(79.13)	79.50
	44.60	*(53.46)*	*(63.85)*	*(72.06)*	*82.28*	*(90.00)*	*94.05*	*98.35*	*(99.54)*	*100.00*
Hanoverian (Stegen, 1929)	34.37	39.74	47.40	52.99	58.35	66.75	68.09	73.39	75.28	76.36
	45.02	*52.04*	*62.07*	*69.40*	*76.41*	*87.41*	*89.17*	*96.11*	*98.59*	*100.00*
Holstein (Iwersen, 1926)	34.31	(41.45)	49.80	54.65	60.10	68.31	71.50	75.62	77.24	78.74
	43.58	*(52.64)*	*63.25*	*69.40*	*76.33*	*86.75*	*90.80*	*96.03*	*98.10*	*100.00*
Quarter Horse (Cunningham and Fowler, 1961)	30.70	(37.71)	45.95	54.55	61.60	68.90	70.05	71.40	73.20	72.65
	41.94	*(51.52)*	*62.77*	*74.52*	*84.15*	*94.12*	*95.70*	*97.54*	*100.00*	*99.25*
Av., above 6 series	34.46	41.21	49.29	56.04	63.44	70.32	73.35	76.70	78.28	78.84
	43.71	*52.28*	*62.52*	*71.09*	*80.47*	*89.19*	*93.04*	*97.29*	*99.29*	*100.00*
Ratios adopted as Optimum	*43.80*	*52.52*	*62.75*	*70.83*	*81.86*	*90.50*	*94.00*	*97.33*	*99.38*	*100.00*

and mares the difference in chest girth amounts, on the average, to only about a quarter inch, the mares being the smaller. This almost equal girth of chest, along with girth of belly, is the main reason why females throughout the entire horse family weigh almost as much, on the average, as males. The lesser weight of the fillies and mares is the result also of more slender necks and limbs, along with generally less massive bones throughout the skeleton. All in all, however, horses and their domestic and/or wild relatives (asses, zebras, onagers) are singularly similar in size between the sexes, in this respect differing markedly from cattle (in which bulls may weigh *twice* as much as cows), and from other forms such as deer, goats, sheep, the larger carnivores (including such marine forms as seals, sea lions, and walruses), anthropoid apes, and, of course, man. In fact, in linear measurements of the bones, horses are only about *half* as variable as humans.

It may be noted that, typically, in all breeds and both sexes, the amount that the chest increases in girth between birth and one year is approximately two-thirds (c. 67 percent) of the ultimate or total amount of gain.

From Table 5 has been developed Table 6, which gives body weight as estimated from chest girth. In only two of the six listings (i.e., Percheron colts and the Quarter Horse) was weight actually included in the original data. Even so, *all* the weights in Table 6 are estimations, which in the two breeds just mentioned correspond closely with the original weights as published. But just as in the case of height (at the withers), which varies from birth to maturity in accordance with the typical or characteristic *size* of the parent animals, so also does body weight. For this reason, as in connection with height, the bottom row in Table 6 should be taken only as a general guide. The writer's formula for birth weight is: .0854 adult weight (of same sex) + 13.15 lbs. This gives the following weights for newborn *males*: Shetland pony, 45 pounds; Arab, 93 pounds; Thoroughbred, 113.5 pounds; draft, 150 pounds. Again, however, there are wide deviations from these theoretical average birth weights in the case of certain reported instances (in which perhaps the foals have been either under- or overfed). For example, the male newborn foals of the Rheinish-Dutch draft horses reported by Hering (1925) have an average weight (as estimated from chest girth) of only 138.6 pounds, or only 7.59 percent of the sire's estimated body weight (1826 pounds). As the other extreme, the female newborns of the Hanoverian breed reported by Stegen (1929) weigh an esti-

TABLE 6

Body Weight (lbs.) as Estimated from the Chest Girths listed in Table 5.
Sexes combined

Breed and Author	Birth	Age								
		1 mo.	3 mos.	6 mos.	1 year	1½ yrs.	2 yrs.	3 yrs.	4 yrs.	5 yrs.
Rheinish-Dutch (Hering, 1925)	136 *8.37*	237 *14.58*	390 *24.00*	584 *35.94*	865 *53.23*	1148 *70.74*	1345 *82.77*	(1451) *(89.29)*	(1586) *(97.60)*	1625 *100.00*
Rheinish-Belgian (Zimmermann, 1933)	172 *9.64*	268 *15.01*	428 *23.98*	626 *35.07*	986 *55.24*	1199 *67.17*	1481 *82.97*	1712 *95.91*	(1766) *(98.94)*	1785 *100.00*
Percheron Colts, Av. of both groups (Trowbridge and Chittenden, 1932)	146 *9.74*	(245) *(16.30)*	(405) *(26.96)*	(573) *(38.16)*	838 *55.80*	(1096) *(72.93)*	1250 *83.17*	1430 *95.15*	(1482) *(98.58)*	1503 *100.00*
Hanoverian (Stegen, 1929)	135 *10.13*	203 *15.22*	333 *25.03*	458 *34.39*	603 *45.30*	889 *66.78*	944 *70.90*	1182 *88.76*	1276 *95.80*	1332 *100.00*
Holstein (Iwersen, 1926)	134 *9.20*	(228) *(15.62)*	384 *26.28*	500 *34.26*	657 *45.00*	953 *65.30*	1093 *74.89*	1293 *88.58*	1378 *94.41*	1460 *100.00*
Quarter Horse (Cunningham and Fowler, 1961)	99 *8.43*	(175) *(14.92)*	305 *26.03*	498 *42.42*	705 *60.10*	978 *83.40*	1028 *87.63*	1089 *92.81*	1173 *100.00*	1147 *97.78*
Av., above 6 series	136 *9.28*	224 *15.32*	373 *25.23*	538 *36.67*	767 *52.36*	1040 *70.96*	1180 *80.53*	1350 *92.08*	1435 *97.89*	1466 *100.00*
Ratios adopted as Optimum	*9.66*	*15.85*	*26.01*	*36.80*	*54.20*	*68.00*	*78.24*	*92.18*	*98.13*	*100.00*

mated 133.3 pounds, which is 10.55 percent of the dam's estimated weight (1264 pounds). It would certainly appear that these variations in birth weights are more likely due to differences in *nutrition* than to peculiarities of the *breed*. That is, the birth weight of the Rheinish-Dutch colts *should* have been at least 160 pounds (rather than 138.6), while that of the Hanoverian fillies *should* have been not over 120 pounds (rather than 133.3).

The point of all this is that the size of the foal at birth, and of the same animal at maturity, along with the rate of growth in between these stages, is bound to vary among different breeds, breeders, and rates of feeding, not to mention among various genetic and hereditary factors. But this does not mean that the breeder cannot benefit by so feeding his pregnant mares and, later, their foals, according to a schedule that lies approximately midway between extremes. For the ideally raised foal is one that is neither underfed nor overstuffed, but an animal in which the degree of growth and development follows that which is common to domestic horses *in general*.

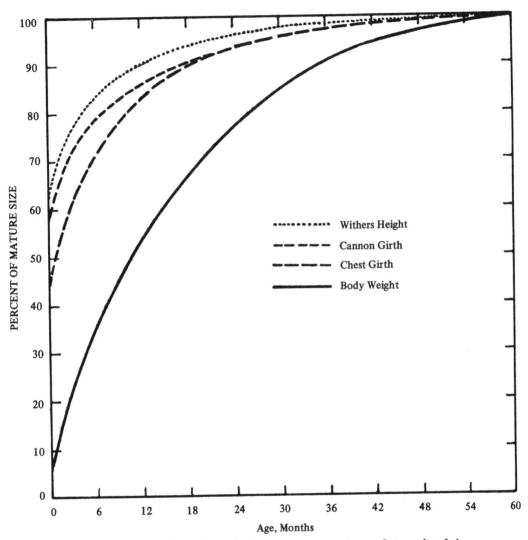

Fig. 2. *Equine growth in four body measurements as determined in this study. For specific ratios, see Tables 3, 4, 5 and 6.*

TABLE 7
Body Weight and Cannon Girth ("Bone") in Relation to Chest Girth

Light Horses

Chest Girth, in.	Weight, lbs. Males	Weight, lbs. Females	Cannon Girth, in. Males	Cannon Girth, in. Females
30	95	90	4.55	4.47
31	104	99	4.63	4.54
32	114	108	4.71	4.61
33	124	118	4.79	4.68
34	134	128	4.87	4.75
35	145	139	4.96	4.83
36	157	150	5.04	4.90
37	169	162	5.12	4.98
38	182	175	5.20	5.05
39	196	188	5.28	5.12
40	210	202	5.37	5.20
41	225	217	5.45	5.27
42	241	233	5.53	5.34
43	258	249	5.61	5.42
44	275	266	5.69	5.49
45	293	283	5.78	5.57
46	311	301	5.86	5.64
47	331	320	5.95	5.72
48	352	340	6.03	5.79
49	373	361	6.11	5.86
50	395	382	6.20	5.94
51	417	404	6.28	6.02
52	440	427	6.36	6.09
53	465	451	6.44	6.16
54	491	476	6.52	6.23
55	517	502	6.61	6.31
56	544	528	6.69	6.39
57	573	556	6.77	6.46
58	602	585	6.85	6.53
59	632	614	6.93	6.61
60	663	644	7.02	6.69
61	694	675	7.10	6.76
62	727	707	7.19	6.84
63	761	740	7.27	6.91
64	797	775	7.35	6.99
65	834	811	7.43	7.06
66	872	848	7.52	7.15
67	911	886	7.63	7.24
68	952	926	7.73	7.33
69	995	968	7.84	7.42
70	1040	1012	7.96	7.52
71	1086	1057	8.07	7.63
72	1133	1103	8.18	7.73
73	1180	1149	8.29	7.83
74	1228	1196	8.40	7.92
75	1278	1244	8.51	8.02
76	1330	1293	8.61	8.11
77	1385	1344	8.72	8.21
78	1440	1397	8.83	8.30

Draft Horses

Chest Girth, in.	Weight, lbs. Males	Weight, lbs. Females	Cannon Girth, in. Males	Cannon Girth, in. Females
30	95	90	5.31	5.23
32	114	108	5.49	5.39
34	134	128	5.67	5.55
36	157	150	5.85	5.71
38	182	175	6.03	5.87
40	210	202	6.21	6.03
42	241	233	6.39	6.19
44	275	266	6.57	6.35
46	311	301	6.75	6.51
48	352	340	6.93	6.67
50	395	382	7.11	6.83
52	440	427	7.29	6.98
54	491	476	7.47	7.14
56	544	528	7.65	7.30
58	602	585	7.83	7.46
60	663	644	8.01	7.62
62	727	707	8.19	7.77
64	797	775	8.37	7.93
66	872	848	8.55	8.09
68	952	926	8.73	8.25
70	1040	1012	8.91	8.41
72	1133	1103	9.09	8.57
74	1228	1196	9.27	8.73
76	1330	1293	9.45	8.89
78	1440	1397	9.63	9.05
80	1553	1510	9.81	9.21
82	1672	1627	9.99	9.36
84	1798	1748	10.17	9.52
86	1929	1876	10.35	9.68
88	2067	2010	10.53	9.84
90	2212	2151	10.71	10.00
92	2363	2298	10.89	10.15
94	2520	2449	11.07	10.31
96	2682	2607	11.25	10.47
98	2853	2774	11.43	10.63
100	3033	2950	11.61	10.79

Shetland Ponies

Chest Girth, in.	Weight, lbs. Males	Weight, lbs. Females	Cannon Girth, in. Males	Cannon Girth, in. Females
20	32	30	3.45	3.44
21	37	34	3.51	3.49
22	42	39	3.58	3.55
23	47	44	3.64	3.61
24	52	49	3.71	3.67
25	58	55	3.78	3.73
26	65	61	3.85	3.79
27	72	68	3.92	3.85
28	79	75	3.99	3.91
29	87	82	4.05	3.97
30	95	90	4.12	4.03
31	104	99	4.18	4.08
32	114	108	4.25	4.14
33	124	118	4.32	4.20
34	134	128	4.39	4.26
35	145	139	4.45	4.31
36	157	150	4.52	4.37
37	169	162	4.59	4.43
38	182	175	4.66	4.49
39	196	188	4.72	4.55
40	210	202	4.79	4.61
41	224	216	4.85	4.66
41	238	230	4.92	4.72
43	253	245	4.99	4.78
44	268	260	5.06	4.84
45	284	276	5.12	4.90
46	301	292	5.19	4.96
47	318	309	5.26	5.02
48	337	327	5.33	5.08
49	357	347	5.39	5.14
50	379	368	5.46	5.20
51	402	391	5.52	5.26
52	426	415	5.59	5.32
53	452	440	5.65	5.38
54	478	465	5.72	5.44
55	505	491	5.79	5.50

Note that for a given girth of chest an adult Shetland has a much smaller cannon than a foal of a light-horse breed; and that similarly the cannons of light horses are much smaller than those of draft horses having the same girth of chest.

3
Growth in the Foal

HORSEMEN ARE ALWAYS INTERESTED IN THE QUESTION OF HOW BIG THEIR foals will be when full grown. In this connection a number of rule-of-thumb methods have been proposed for estimating from the foal the probable height at the withers when full growth has been attained. One such method is to measure (with a tape) the height from the bottom of the front fetlock to the "bottom" of the elbow, and from the latter point over the back to the highest midpoint on the withers. In an adult horse these two measurements are approximately equal. However, any error in the taking of the lower (fetlock to elbow) measurement will be doubled in applying it to the upper (elbow to withers) measurement, and such a degree of error will make the predicted withers height useless as a guide. Moreover, it is difficult, if not impossible, to locate precisely and uniformly such an indefinite point as the "bottom", or lower border, of the elbow (olecranon process of the ulna), even if the measuring is done by an expert.

Figure 3, which is based on the Arabian horse but also applies fairly accurately to all breeds so far as *proportionate* growth is concerned, depicts the lengths of the principal bones of the limbs in relation to age. While, as will be seen, the metacarpus (front cannon bone) reaches essentially its full length by the end of the first year, the other long bones (including the scapula, which is not charted) continue to lengthen until about the age of four years. This means that total "leg length," or the height from the ground to the top of the elbow, normally increases until the age of about four years, even though the increase amounts to only about 1¼–1½ inches after the age of one year and is due mainly to lengthening of the radius or principal forearm bone.

However, even after all increase in leg length (elbow height) has taken place, the height at withers normally increases by about eleven percent (see bottom row of Table 3) from one year to four years. This percentage increase in a horse of average size amounts

49

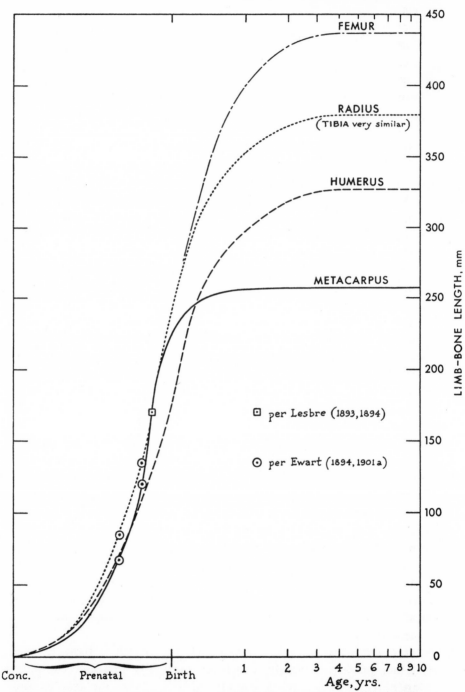

Fig. 3. *Growth in length of the limb bones in the Arabian horse.*

to a gain of over *six inches* in withers height. This gain is due wholly to increase in the girth of the chest (about 15 inches) and consequently to increase in vertical *chest depth* (from sternum to withers). That is, the increase in chest *depth* amounts to about 40 percent of the increase in chest *girth*. This proportionate increase is confirmed by a comparison of the figures at one year and at four years in Tables 3 and 5. A slight additional gain in chest girth and consequently in chest depth and withers height is made after the age of four years.

To serious breeders, I strongly recommend that the physical growth in their foals be gauged by actual figures such as the foregoing. For it is the custom of certain writers on horses to quote some ancient Arab "rule" of growth which by its very naiveté does away with all possibility of accuracy.

Table 8, which gives optimum body weights at later stages compared with that at birth, has been derived from the figures and ratios presented in Table 6. Accordingly, the weights listed in Table 8 represent good, average values that are neither too light nor too heavy for foals of the birth weights commenced with. As has been pointed out by Brody and others, optimum or superior growth is not maximum. By *superior* is meant those animals which are the most efficient users of food and have the highest vitality. Contrary to common belief, the most rapid growth is often associated with the highest early mortality. "Very rapid growth may be economical to the animal husbandman by saving overhead maintenance cost but not profitable to the animal whose longevity may be impaired thereby" (Brody, 1964, p. 9). That many a foal has doubled its weight during the first month is, therefore, no basis for assuming that all foals (or their dams) should be fed with such a gain in view. Often a foal that has been undernourished in its prenatal stage will make such a gain, but this is simply an instance of *compensatory* growth. Wherever possible, it is best that the young animal's growth in weight should follow the average or typical trend, rather than fluctuate up and down throughout the entire development period. However, where the main source of nourishment is pasturage, such an ideal can hardly be followed if the food supply is subject to wide variations.

It may be mentioned that dairy cattle, from birth to one month of age, gain on the average only about 10 or 11 percent in chest girth and 31–32 percent in weight (Brody, 1964, p. 653). In horses of comparable size, the weight gain during the same period, as shown

TABLE 8

Optimum Body Weight at Various Stages of Growth in Relation to Birth Weight*

(weights apply to *both* sexes)

Birth Weight		Bodyweight, lbs., at age of:									
Lbs.	Percent of Adult Weight	1 mo.	3 mos.	6 mos.	1 yr.	1½ yrs.	2 yrs.	3 yrs.	4 yrs.	5 yrs.
30	15.20	41	60	80	112	138	157	183	194	197
35	13.60	50	75	101	143	177	202	237	251	256
40	12.72	59	90	122	175	217	248	291	309	314
45	12.06	68	105	144	206	256	294	344	366	373
50	11.59	76	119	165	238	296	340	398	424	432
55	11.22	84	133	186	269	336	385	452	481	490
60	10.94	93	148	207	301	376	431	506	539	549
65	10.71	102	163	228	332	415	476	560	596	607
70	10.52	111	178	249	364	455	522	614	653	666
75	10.36	119	192	270	395	494	568	668	710	724
80	10.22	128	207	291	426	534	614	722	768	783
85	10.10	137	222	312	457	573	659	776	825	841
90	10.00	146	237	333	489	613	705	830	883	900
95	9.91	154	251	354	520	652	750	884	940	958
100	9.83	163	266	375	552	692	796	938	998	1017
105	9.76	171	280	396	583	731	841	991	1055	1075
110	9.70	180	295	418	615	771	887	1045	1113	1134
115	9.64	189	310	439	646	810	933	1099	1170	1192
120	9.59	198	325	460	678	850	979	1153	1228	1251
125	9.54	206	339	481	709	890	1024	1207	1285	1309
130	9.50	215	354	502	741	930	1070	1261	1343	1368
135	9.46	223	369	523	772	969	1115	1315	1400	1426
140	9.42	232	384	544	803	1009	1161	1369	1458	1485
145	9.39	241	398	565	834	1048	1207	1423	1515	1543
150	9.36	250	413	586	866	1088	1253	1477	1572	1602
155	9.33	258	427	607	897	1127	1298	1531	1630	1661
160	9.30	267	442	629	929	1167	1344	1585	1688	1720
165	9.28	275	456	650	960	1206	1389	1638	1745	1778
170	9.26	284	471	671	992	1246	1435	1692	1802	1837
175	9.23	293	486	692	1023	1285	1481	1746	1859	1895
180	9.21	302	501	713	1054	1325	1527	1800	1917	1954
185	9.19	310	516	734	1084	1365	1573	1854	1974	2013

Shetlands: rows 30–50
Larger Ponies: rows 55–85
Light Horses: rows 90–130
Medium to Heavy Draft Horses: rows 135–185

* If the foal's weight is not known, or is unobtainable, *estimate* it from chest girth as per Table 7.

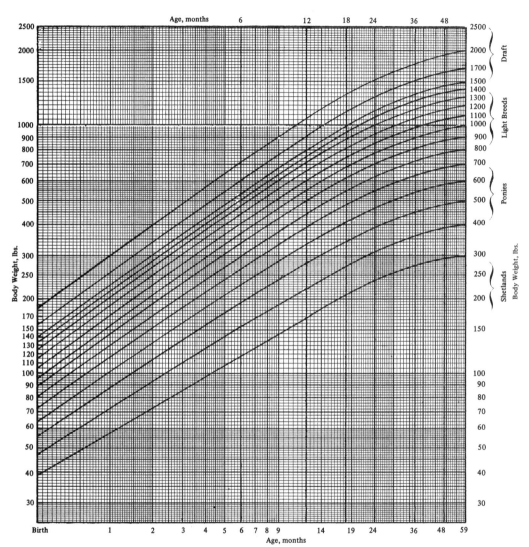

Fig 4.   Optimum Body Weight in Relation to Age
When a foal's weight at successive ages is plotted on this graph,
the more closely the weights follow the trend of the growth lines,
the more normal the rate of growth. A zigzag line, on the contrary, indicates erratic growth and feeding.

in Table 8, averages about 60 percent, or practically double the gain
in the cows. As a consequence of the horses' rate of growth, it develops that the body weight at three months (for all sizes, from
Shetlands to drafters) is approximately halfway between that at
birth and at six months. From six months to eighteen months the
gain, although gradually diminishing, is more uniform, so that the
weight at nine months is only a trifle over halfway between six and
twelve months; and at fifteen months only a trifle over halfway between twelve and eighteen months.

Figure 4 is a graphic charting of the bodyweight-age relationships
listed in Table 8. By using this chart, the weight of a foal may be
plotted on it and the animal's progress in size clearly recorded. The
more nearly the weight follows the course of a given line on the
graph, the more it will conform with the optimum rate of growth.
Table 8 and Figure 4, used in conjunction, should eliminate the
need for any empirical means of predicting the best course of growth
in the newborn or weanling foal of either sex.

4
Detailed Body Measurements in Certain Popular Breeds

WHILE, AS MENTIONED, WEIGHT IS THE BEST SINGLE MEASURE OF BODILY size and index of growth in the foal, it is desirable also to know as far as possible the entire physical makeup of the horse in its principal breeds and types, and what exact measurements express the differences in those breeds. I have accumulated sufficient data to enable such measurements to be presented in mature stallions and mares (i.e., at five years of age) in the following breeds: Arab, Thoroughbred, Standardbred, Morgan, Appaloosa, Quarter Horse, Shetland pony, and draft horse.* In addition, in the Arab, Thoroughbred, Quarter Horse, Shetland, and draft, respectively, comparative measurements are given for newborn foals. These breeds will now be discussed in the order just listed.

Arab

For at least the last hundred years, and probably much longer, there has been no significant change in the physique of the pure-bred Arabian horse. In withers height desert-bred individuals have averaged 14-2 hands (58 inches) and European-bred specimens 14-3 hands or a trifle over. The *range* in height, however, may vary between 54 and 64 inches. While the typical height of Arabs has been stated in innumerable books and articles, seldom have any other measurements, even body weight, been given. A perusal of the literature shows that the native breeders in Arabia and North Africa, who before the advent of motorized transportation used their horses mainly in warfare or pillaging, demanded lean, hardy animals pos-

---

* Wherever a specific measurement differs from that given in my previous books, that given here is to be given preference, as being based on more extensive data.

55

sessed of the utmost attainable speed and endurance. As a consequence, such warriors abhorred a "fat," complacent horse and fairly worshipped one on which they could rely in any emergency. In warfare it was customary to use mares rather than stallions, for the asserted reason that in the dark stallions were more likely to neigh and thus reveal the presence of their masters to the enemy.

As an insurance of speed, therefore, the Arab should be built along the lines of a Thoroughbred, which in fact he is. And in order to be proportioned like a three-year-old racehorse weighing, say, 1090 pounds, an adult Arab stallion, at its lesser height, should weigh about 933 pounds. This theoretical ideal weight is confirmed by that recorded for well-bred European Arabs that have not been overfed and have had adequate exercise. The following measurements of adult Arabs are based mainly on the data of Madroff (q.v., Tables 1 and 2) and myself, respectively, since from my own measurings I have determined also the average body proportions of newborn foals. For a diagram showing where the customary measurements are taken, see Figure 1.

Some of the measurements in Table 9, such as the heights of the knee, elbow, and hock, were established not only from measurements taken on living horses, but also from the lengths of the limb bones in six male and two female Arab skeletons I have studied. Similarly, the length and width of the head were checked by the typical proportions shown in 24 Arab skulls. Neck length was based both on that measured in living animals and on the lengths of the individual neck vertebrae measured in several skeletons. From these exact figures, a number of traditional beliefs concerning the supposedly different body proportions and constitution of the Arab horse (e.g., "long neck," "high withers," "very broad forehead," "short cannons," "dense bone," etc.) are shown, by a comparison of these characters with those existing in other light-horse breeds, to have no basis in fact. Such minor differences in body conformation as do occur between the Arab and its chief descendant, the Thoroughbred, are pointed out in the following discussion of the latter breed. In making comparisons, the figures in the second columns, which indicate *proportions,* are more significant than those in the first columns, which show absolute (average) *size.*

Whatever the bodily conformation of the Arabian horse, the important point to remember is that his value as a contributor of desirable qualities in cross-breeding has been, and still is, inestimable.

TABLE 9

Measurements and Proportions of the typical Purebred Arab*
(% = Percent of Withers Height)

	Measurement	Adult Males		Adult Females		Newborn Males		Newborn Females	
		in.	%	in.	%	in.	%	in.	%
A	Height, withers	59.70	100.00	59.00	100.00	37.50	100.00	37.06	100.00
B	Height, croup	59.40	99.50	58.82	99.70	38.29	102.10	37.86	102.15
C	Length, trunk (slantwise)	59.28	99.30	58.88	99.80	28.50	76.00	28.22	76.15
D	Length, head	23.34	39.10	22.83	38.70	13.84	36.90	13.62	36.75
E'	Width, forehead (max.)	8.95	15.00	8.75	14.83	5.48	14.60	5.39	14.54
F	Length, ear (externally)	7.16	12.00	7.01	11.88	5.66	15.10	5.57	15.03
G	Length, neck	27.52	46.10	27.33	46.32	12.45	33.20	12.32	33.24
H	Girth, neck (at throat)	29.43	49.30	27.73	47.00	15.60	41.60	15.44	41.66
J	Girth, chest (max.) (behind withers)	67.51	113.08	66.96	113.50	29.74	79.30	29.44	79.44
K'	Width, shoulders (max.)	15.40	25.80	14.93	25.30	7.88	21.00	7.81	21.06
L	Height, chest from ground	32.70	54.80	32.20	54.59	23.78	63.40	23.48	63.36
M	Height, elbow	35.70	59.80	35.46	60.10	25.88	69.00	25.73	69.43
N	Height, knee (top of pisiform)	18.63	31.20	18.38	31.15	14.63	39.00	14.54	39.24
P	Girth, forearm	18.62	31.19	17.94	30.41	11.07	29.51	10.84	29.25
Q	Length, back	26.92	45.10	26.67	45.20	13.31	35.50	13.18	35.57
R^P	Length, croup (projected)	19.52	32.70	19.29	32.70	8.89	23.70	8.82	23.80
S'	Width, croup (max.)	19.82	33.20	20.24	34.30	8.63	23.00	8.85	23.88
T^P	Length, hip to hock (projected)	34.86	58.40	34.60	58.64	20.32	54.20	20.09	54.21
U	Girth, gaskin (max.)	16.13	27.02	15.69	26.59	9.73	25.95	9.53	25.71
V	Girth, fore cannon (min.)	7.36	12.33	7.12	12.08	4.40	11.73	4.32	11.66
W	Girth, hind cannon (min.)	8.01	13.42	7.82	13.26	4.71	12.55	4.62	12.48
X	Height, hock	23.94	40.10	23.62	40.04	18.75	50.00	18.53	50.00
Y	Length, fore hoof (sole)	5.91	9.90	5.37	9.10	2.89	7.70	2.83	7.63
Y'	Length, hind hoof (sole)	5.77	9.70	5.25	8.90	3.08	8.20	3.01	8.14
Z	Width, fore hoof	4.72	7.90	4.42	7.50	2.35	6.27	2.30	6.22
Z'	Width, hind hoof	4.24	7.10	3.95	6.70	2.35	6.27	2.30	6.22
Body Weight, lbs.		933		881		93		90	
Body Build		0.845		0.835		0.345		0.345	

* For withers heights other than the average figures given in this table, use the percentage ratios to obtain the corresponding measurements in inches. Body weight may be estimated by the chest girth formulas given on page 38.

Fig. 5. Relative size and body proportions of an Arab mare and new-born foal. Compare with Figs. 12 and 14. x .06 natural size.

Thoroughbred

So much has been written about the history of the Thoroughbred that it should be unnecessary to deal with the subject here.* It must be evident, from a comparison of the physiques of the Thoroughbred and the Arab (see Tables 9 and 10), that the former is essentially a larger counterpart of the latter, modified only insofar as the development of greater running speed is concerned. However, in referring to the ancestry of the Thoroughbred, it would be more cor-

---

* For a good early reference work on the history of the English Thoroughbred, see Youatt, *The Horse* (1880) , pp. 54–66.

rect to use the term *Oriental horse* (or stock) than to confine the relationship to the Arab, since the early-day progenitors of the breed included, besides Arabs, many outstanding Barbs, Turks, Persians, Syrians, and other Near East types, which in many cases were so alike Arabs as to be confused with them, even by their owners.

Secondly, whatever the ancestry of the English mares to which the imported stallions were mated, it may safely be said that the Thoroughbred of today, whether bred in England, on the Continent, in Australia, the United States, or elsewhere, is one hundred percent "warmblood." And even if some of the foundation sires were "cart horses" (e.g., the Godolphin Barb), it does not necessarily mean that they were "coldbloods," since horses of the light, Oriental type may be of large size (say, up to 1400 pounds) and still have no draft blood in their makeup.

In comparing the body proportions (not the absolute measurements, but the *proportions*) of the Thoroughbred with those of the Arab (see Tables 9 and 10) about the only significant difference is the longer neck of the Thoroughbred. Both breeds have relatively small heads, short trunks, and long, slender limbs. Indeed, there is no basis for saying, as some authors insist on doing, that the cannons of the Arab should be "short," when in reality they are essentially as long as in the Thoroughbred. That long, not short, cannons are correlated with running speed is shown by the exceedingly long cannons characteristic of the Asiatic equids known as *half-asses* (onagers, kulans, and kiangs), which animals, in proportion to their size, are swifter than any trained racehorse, along with having remarkable endurance.*

Figure 6 depicts an "ideal" Thoroughbred stallion, drawn by the writer after a great deal of study both of living horses and horse skeletons. Although the drawing was made years before the figures in Table 10 were established, it shows essentially the same proportions of the body, along with exact measurements and normal inclinations of the skull, vertebrae, pelvis, and limb bones. The source of most of the measurements incorporated in this drawing and in Table 10 are from Plischke (1927). A few useful measurements of

---

* According to Roy Chapman Andrews, who headed an expedition into the desert of Dzungaria (Mongolia), a kulan stallion pursued on the open plain was run down only after 29 miles, during 16 miles of which he averaged 30 mph (Andrews, 1933, p. 7).

TABLE 10
Measurements and Proportions of the typical Thoroughbred*
(% = Percent of Withers Height)

	Measurement	Adult Males		Adult Females		Newborn Males		Newborn Females	
		in.	%	in.	%	in.	%	in.	%
A	Height, withers	63.78	100.00	62.99	100.00	39.83	100.00	39.34	100.00
B	Height, croup	63.50	99.57	63.33	100.54	40.67	102.10	40.19	102.15
C	Length, trunk (slantwise)	63.78	100.00	63.47	100.76	30.27	76.00	29.96	76.15
D	Length, head	24.52	38.45	23.97	38.05	14.70	36.90	14.46	36.75
E'	Width, forehead (max.)	9.38	14.70	9.17	14.55	5.82	14.60	5.72	14.54
F	Length, ear (externally)	7.51	11.78	7.35	11.66	6.01	15.10	5.91	15.03
G	Length, neck	31.16	48.86	30.89	49.03	13.22	33.20	13.08	33.24
H	Girth, neck at throat)	31.94	50.08	29.12	46.24	16.70	41.93	16.57	42.12
J	Girth, chest (max.) (behind withers)	72.90	114.30	72.45	115.02	31.85	79.96	31.59	80.30
K'	Width, shoulders (max.)	16.49	25.85	15.94	25.30	8.44	21.19	8.38	21.30
L	Height, chest from ground	35.00	54.88	34.37	54.55	25.14	63.12	24.77	62.96
M	Height, elbow	38.32	60.08	38.07	60.43	27.37	68.72	27.16	69.03
N	Height, knee (top of pisiform)	20.12	31.55	19.84	31.50	15.47	38.84	15.35	39.02
P	Girth, forearm	19.72	30.92	18.91	30.02	11.86	29.77	11.63	29.56
Q	Length, back	29.27	45.90	28.98	46.00	14.14	35.50	13.99	35.57
R^P	Length, croup (projected)	20.83	32.07	20.67	32.81	9.44	23.70	9.36	23.80
S'	Width, croup (max.)	21.37	33.50	21.75	34.52	9.16	23.00	9.39	23.88
T^P	Length, hip to hock (projected)	37.88	59.40	37.58	59.65	21.59	54.20	21.32	54.20
U	Girth, gaskin (max.)	17.54	27.50	16.96	26.92	10.39	26.09	10.21	25.95
V	Girth, fore cannon (min.)	8.10	12.70	7.64	12.13	4.71	11.83	4.62	11.74
W	Girth, hind cannon (min.)	8.82	13.80	8.50	13.49	5.04	12.66	4.96	12.60
X	Height, hock	25.45	39.90	25.14	39.90	19.91	50.00	19.67	50.00
Y	Length, fore hoof (sole)	6.31	9.90	5.80	9.20	3.07	7.70	3.00	7.63
Y'	Length, hind hoof (sole)	6.00	9.40	5.54	8.80	3.27	8.20	3.20	8.14
Z	Width, fore hoof	5.10	8.00	4.54	7.20	2.49	6.27	2.45	6.22
Z'	Width, hind hoof	4.59	7.20	4.22	6.70	2.49	6.27	2.45	6.22
	Body Weight, lbs.	1175		1114		113.5		110	
	Body Build	0.888		0.867		0.352		0.352	

* For withers heights other than the average figures given in this table, use the percentage
ratios to obtain the corresponding measurements in inches. Body weight may be estimated
by the chest girth formulas given on page 38.

English Thoroughbreds are given also by von Nathusius (1905).* The typical height here adopted for Thoroughbred stallions—63.78 inches—is the English equivalent of the metric height of 1620 millimeters given by Plischke. Similarly, the 62.99 inches here listed for mares is equivalent to Plischke's 1600 mm.

Standardbred

The trotting and pacing horses used in harness racing are essentially the same in body size and proportions as the light carriage horses known as "roadsters." Both are of the breed called "Standardbred." Table 11 lists the average or typical body measurements in this breed as determined by me from the data available. These data have consisted mainly of the measurements taken by Grange (1894) of eleven Roadsters, along with similar dimensions listed by Hervey (1941) of six champion trotting stallions. Certain measurements of these Standardbred horses were checked also against those given by Afanassieff (1930) for Orloff trotters, this Russian breed being very similar in its bodily makeup to that of American harness horses, although somewhat smaller in absolute dimensions (height, weight, etc.).

In their general physical conformation, Standardbred horses are essentially smaller editions of Thoroughbreds. Such slight differences as occur between the two breeds may be attributed to the geometric effects of smaller size in the Standardbred, along with minor changes brought about by the differing gaits (galloping vs. trotting or pacing) characteristic of the two breeds.

Newborn Standardbred colts average 39.38 inches in withers height and about 106 pounds in weight, with fillies being slightly smaller.

Quarter Horse

This highly popular Western stock horse of Colonial origin** has a registration that exceeds in number that of all other breeds in

---

* So far as "flat" measurements (i.e., widths, lengths, and depths) are concerned, a valuable table dealing with various breeds is given in the German edition (but not in the English reprint by Dover publications) of *An Atlas of Animal Anatomy*, by W. Ellenberger, H. Baum, and H. Dittrich (1949). The table lists no fewer than 65 different external measurements in 18 different horses, including an English Thoroughbred stallion and mare.

** The Quarter Horse is said to be a cross between Chickasaw stallions and English Thoroughbred mares, the Chickasaw horse in turn having been bred by Indians from Spanish Barbs, Arabs, and Turks brought to America by the conquistadors. The first crossings were made about 1611.

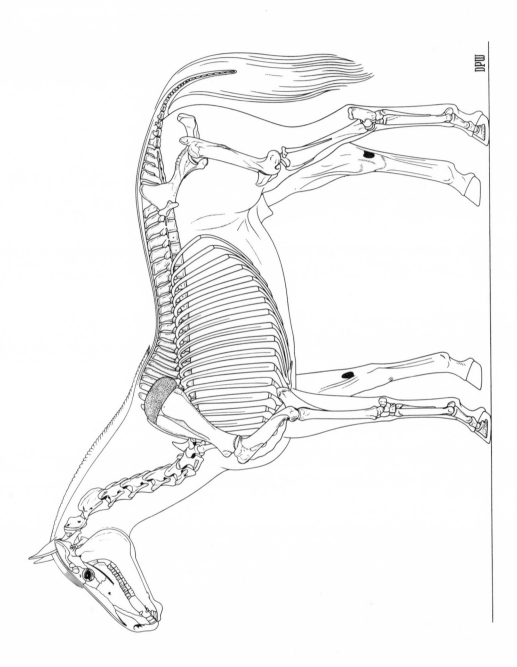

Fig. 6. A drawing by the author, to exact scale, of an "ideal" Thoroughbred stallion.   x .06 natural size.

TABLE 11
Measurements and Proportions of the typical Standardbred*
(% = Percent of Withers Height)

	Measurement	Adult Males		Adult Females	
		in.	%	in.	%
A	Height, withers	63.00	100.00	62.22	100.00
B	Height, croup	62.81	99.70	62.03	99.70
C	Length, trunk (slantwise)	63.10	100.16	62.87	101.05
D	Length, head	25.01	39.70	24.54	39.44
E'	Width, forehead (max.)	9.58	15.20	9.36	15.04
F	Length, ear (externally)	7.62	12.10	7.45	11.98
G	Length, neck	29.61	47.00	29.40	47.25
H	Girth, neck (at throat)	31.05	49.29	28.62	46.00
J	Girth, chest (max.) (behind withers)	71.00	112.70	70.50	113.30
K'	Width, shoulders (max.)	16.76	26.60	15.80	25.40
L	Height, chest from ground	34.60	54.76	34.02	54.68
M	Height, elbow	37.67	59.80	37.42	60.14
N	Height, knee (top of pisiform)	19.66	31.20	19.39	31.17
P	Girth, forearm	19.30	30.63	18.50	29.73
Q	Length, back	28.91	45.89	28.63	46.01
Rᴾ	Length, croup (projected)	20.64	32.76	20.49	32.93
S'	Width, croup (max.)	21.02	33.40	21.47	34.50
Tᴾ	Length, hip to hock (projected)	37.99	60.30	37.69	60.58
U	Girth, gaskin (max.)	17.17	27.25	16.59	26.66
V	Girth, fore cannon (min.)	7.92	12.57	7.50	12.06
W	Girth, hind cannon (min.)	8.56	13.59	8.35	13.42
X	Height, hock	24.95	39.60	24.64	39.60
Y	Length, fore hoof (sole)	6.11	9.70	5.60	9.00
Y'	Length, hind hoof (sole)	6.11	9.70	5.60	9.00
Z	Width, fore hoof	4.98	7.90	4.67	7.50
Z'	Width, hind hoof	4.47	7.10	4.17	6.70
	Body Weight, lbs.	1085		1033	
	Body Build	0.849		0.835	

* For withers heights other than the average figures given in this table, use the percentage ratios to obtain the corresponding measurements in inches. Body weight may be estimated by the chest girth formulas given on page 38.

Fig. 7. A Standardbred harness horse (pacer) in action. Painting by
John Mariani, courtesy Mrs. John Mariani.

the United States. Yet physically the Quarter Horse is somewhat of
an anomaly. For although it has a build practically as heavy as that
of a draft horse, it nevertheless excels in short-distance racing, with
a speed equal to that of the fastest, slender-built Thoroughbred. The
quickness and agility of the Quarter Horse are indeed the reasons
why it is so popular and efficient as a cow pony, cutting horse, barrel racer, calf roper, and general utility ranch horse. The calm disposition characteristic of this breed makes it also an appropriate
Western pleasure horse, especially for women and junior riders.

The relatively low height, long body, deep chest, and consequently
shorter legs of the Quarter Horse correspond to the physical features
of a work horse rather than a sprinter. Yet by sheer muscular
power this specialized example of equine development is able to
gallop with astonishing speed. In this respect the Quarter Horse is
comparable to certain light-heavyweight human sprinters who can

TABLE 12
Measurements and Proportions of the typical Quarter Horse*
(% = Percent of Withers Height)

	Measurement	Adult Males		Adult Females		Newborn Males		Newborn Females	
		in.	%	in.	%	in.	%	in.	%
A	Height, withers	59.30	100.00	58.50	100.00	37.28	100.00	36.83	100.00
B	Height, croup	59.06	99.60	58.32	99.70	38.07	102.14	37.64	102.20
C	Length, trunk (slantwise)	63.25	106.67	62.93	107.57	28.33	76.00	28.05	76.15
D	Length, head	23.48	39.60	23.48	40.13	13.76	36.90	13.54	36.75
E'	Width, forehead (max.)	8.80	14.84	8.80	15.04	5.44	14.60	5.36	14.54
F	Length, ear (externally)	7.12	12.00	7.12	12.17	5.63	15.10	5.54	15.03
G	Length, neck	28.76	48.50	28.10	48.03	12.38	33.20	12.24	33.24
H	Girth, neck (at throat)	32.26	54.40	30.25	51.71	16.85	45.20	16.72	45.39
J	Girth, chest (max.) (behind withers)	73.65	124.20	72.95	124.70	32.05	85.97	31.70	86.07
K'	Width, shoulders (max.)	17.09	28.82	16.52	28.24	8.49	22.75	8.42	22.87
L	Height, chest from ground	31.72	53.50	31.15	53.25	23.65	63.45	23.35	63.40
M	Height, elbow	34.99	59.00	34.75	59.40	25.65	68.80	25.45	69.10
N	Height, knee (top of pisiform)	18.32	30.89	18.04	30.84	14.52	38.90	14.40	39.10
P	Girth, forearm	19.98	33.69	19.21	32.84	12.40	33.26	12.17	33.04
Q	Length, back	28.73	48.45	28.63	48.94	13.23	35.50	13.10	35.57
R^P	Length, croup (projected)	20.66	34.84	20.51	35.06	8.84	23.70	8.76	23.80
S'	Width, croup (max.)	21.80	36.76	22.20	37.95	8.61	23.10	8.82	23.95
T^P	Length, hip to hock (projected)	35.82	60.40	35.54	60.75	20.20	54.18	20.10	54.58
U	Girth, gaskin (max.)	17.79	30.00	17.18	29.37	10.46	28.06	10.27	27.88
V	Girth, fore cannon (min.)	7.85	13.24	7.40	12.65	4.71	12.63	4.62	12.54
W	Girth, hind cannon (min.)	8.49	14.32	8.21	14.03	5.04	13.52	4.96	13.46
X	Height, hock	23.29	39.27	22.97	39.27	18.64	50.00	18.41	50.00
Y	Length, fore hoof (sole)	6.05	10.20	5.56	9.50	3.08	8.26	3.00	8.15
Y'	Length, hind hoof (sole)	5.93	10.00	5.44	9.30	3.28	8.80	3.20	8.69
Z	Width, fore hoof	4.92	8.30	4.56	7.80	2.50	6.57	2.45	6.65
Z'	Width, hind hoof	4.35	7.50	4.10	7.00	2.50	6.57	2.45	6.65
	Body Weight, lbs.	1178		1117		108		104	
	Body Build	1.097		1.086		0.409		0.405	

* For withers heights other than the average figures given in this table, use the percentage ratios to obtain the corresponding measurements in inches. Body weight may be estimated by the chest girth formulas given on page 38.

run 100 or 220 yards as fast as any competitors of slighter and more typical physique for these events.

As the figures in Table 12 show, the Quarter Horse is of relatively heavy body build (about 1.09, as compared with 0.84 in the Arab and 0.88 in the Thoroughbred). In general girth relative to height, he is about 9 percent larger than the lighter-built Arab and Thoroughbred. This is seen especially in the girths of the chest, forearms, and gaskins. In "bone" (cannon girth), however, the Quarter Horse is rather slender in relation to his weight, although for his height this measurement exceeds that of both the Arab and the Thoroughbred.

It would appear that the Quarter Horse of today is a *developed* example of a short-distance runner, his relatively large muscles having resulted from the vigorous efforts of short duration that similarly would develop the leg muscles of a human sprinter.

My conclusions on the size and body proportions of the Quarter Horse are based mainly, although with numerous exceptions, on the measurements listed in the monograph on this breed by Cunningham and Fowler (1961). For instance, the average height of stallions as given in Table 12 is that of fourteen well-known Quarter Horse sires *in addition* to the thirteen adult males (which average 60 inches in height) given in Cunningham and Fowler's Table 1. These combined data yield an average withers height of 59.3 inches, which corresponds more typically with the latter authors' average height in sixty mares of 58.5 inches. Other differences occur all through Table 12, many of which are due to differences in measuring techniques.

In several articles by various authors on the conformation of the Quarter Horse, numerous unsupported statements have been made. One is that the Quarter Horse is so heavily built in his forequarters that he must support "from one-third to one-half more weight per hand on his forelegs than the Thoroughbred." As will be seen from Tables 10 and 12, the typical Thoroughbred stallion weighs 1175/63.78, or 18.42 pounds per inch of height, while the typical Quarter Horse stallion weighs 1178/59.30, or 19.09 pounds per inch of height. This is a difference of about 3.6 percent for the entire body, and it is highly unlikely that the weight of the forehand of the Quarter Horse would increase this to a total of more than 4 or 5 percent. This is a far cry from "one-third to one-half" more weight than in the Thoroughbred.

In another account, it is remarked, "A cannon needs no length,

Fig. 8. A painting by Orren Mixer adopted by the American Quarter Horse Association as representing an "ideal" stallion of this breed. Courtesy American Quarter Horse Association.

for its sole purpose is support." (The super-swift, long-cannoned wild equines or kulans of central Asia would surely be glad to know this!) In the same article, an illustration shows the shoulder blade of a Quarter Horse inclined at an angle of only 45 degrees from the horizontal, while the upper arm is inclined at an even more impossible angle of only 30 degrees! Fallacies such as these, and their unfortunate publication by editors who should know better, certainly indicate the need for more information on equine conformation based on fact (rather than supposition).

Morgan

The history of the Morgan horse is unique in that the breed originated in a single progenitor, a stallion named Justin Morgan.* While the Morgan horse is in some quarters referred to as the oldest American breed, its foundation sire Justin Morgan was not foaled until 1789, whereas the Quarter Horse, as a distinct breed, originated at least as early as 1752 with the importation of the sire Janus. And quarter-mile racing took place in Virginia as early as 1674. The first "Quarter Horses" were developed at an even earlier date, when Chickasaw stallions bred by American Indians were crossed with imported English mares (see Quarter Horse). Like the Quarter Horse, the Morgan is predominantly of Barb, Arab, and Thoroughbred blood, although some investigators claim Dutch horse ancestry also. The latter might account for the relatively longer body of the Morgan horse—a proportion which definitely is not characteristic of the lighter breeds just mentioned. One of the distinguishing features of Morgan horses is that they are great roadsters or trotters, and in the latter capacity have contributed significantly to the development of the Standardbred racehorse.

The first Morgan horse—Justin Morgan—has been variously described, mostly in superlatives. While most accounts agree that he stood about 14 hands (56 inches), his weight has been given as anywhere from 800 to "never . . . much over a thousand pounds." Since an average-sized Morgan stallion stands 60 inches at the withers, yet weighs only slightly over 1000 pounds, it is evident that Justin Morgan could not have weighed that much at a height four inches less and still have been of typical Morgan horse build. If, at 56 inches height, he *had* been of typical build for his breed, he would have weighed about 840–850 pounds. The heaviest-built stallion, among 39 in a series listed by D. C. Linsley (1857), weighed 875 pounds at a height of 55 inches; while the lightest-built weighed 882 pounds at a height of 61 inches. The average height among these 39 stallions was 60 inches and the average weight 1035 pounds. These figures I have used as a basis for some of the measurements listed in Table 13. Other more detailed measurements, of eight stallions and ten mares, were furnished to me in 1948 by W. M. Daw-

---

* As a young colt, this horse was named Figure; but later on, in honor of the owner who brought the colt from West Springfield, Massachusetts to Vermont, where it became famous, the animal's name was changed to Justin Morgan.

Fig. 9. An artistic conception of a typical Morgan horse. Courtesy the American Morgan Horse Association, Inc.

son, Animal Husbandman of the Bureau of Animal Industry, Washington, D.C.

The general body build and proportions of the Morgan horse, as shown in Table 13, lie partway between those of such light horses as the Arab, Thoroughbred, and Standardbred, and the significantly heavier-built Quarter Horse. Many an illustration depicting the Morgan shows a physique hardly distinguishable from that of an Arab except for a slightly longer body, thicker neck, and a less-refined head carried more erectly.

For all-around usefulness and popularity, the Morgan ranks high, along with the Arab, the Quarter Horse, and the Appaloosa, the latter of which we shall consider next.

TABLE 13
Measurements and Proportions of the typical Morgan*
(% = Percent of Withers Height)

	Measurement	Adult Males		Adult Females	
		in.	%	in.	%
A	Height, withers	60.00	100.00	59.30	100.00
B	Height, croup	59.72	99.54	59.42	100.20
C	Length, trunk (slantwise)	62.28	103.80	61.90	104.38
D	Length, head	23.77	39.62	23.34	39.36
E'	Width, forehead (max.)	9.11	15.17	8.90	15.00
F	Length, ear (externally)	7.25	12.08	7.09	11.96
G	Length, neck	29.16	48.60	28.90	48.73
H	Girth, neck (at throat)	30.92	51.53	28.55	48.15
J	Girth, chest (max.) (behind withers)	69.11	115.18	68.67	115.80
K'	Width, shoulders (max.)	16.04	26.73	15.53	26.19
L	Height, chest from ground	32.28	53.80	31.87	53.75
M	Height, elbow	35.58	59.30	35.28	59.50
N	Height, knee (top of pisiform)	18.61	31.02	18.32	30.90
P	Girth, forearm	19.29	32.15	18.51	31.21
Q	Length, back	28.30	47.17	28.13	47.44
R^P	Length, croup (projected)	20.38	33.97	20.20	34.06
S'	Width, croup (max.)	20.73	34.55	21.13	35.63
T^P	Length, hip to hock (projected)	35.28	58.80	35.02	59.05
U	Girth, gaskin (max.)	16.96	28.27	16.44	27.72
V	Girth, fore cannon (min.)	7.80	13.00	7.36	12.42
W	Girth, hind cannon (min.)	8.49	14.15	8.21	13.84
X	Height, hock	23.90	39.83	23.62	39.83
Y	Length, fore hoof (sole)	5.94	9.90	5.40	9.10
Y'	Length, hind hoof (sole)	5.80	9.67	5.28	8.90
Z	Width, fore hoof	4.77	7.95	4.47	7.54
Z'	Width, hind hoof	4.28	7.13	3.99	6.73
Body Weight, lbs.		1035		890	
Body Build		0.940		0.915	

* For withers heights other than the average figures given in this table, use the percentage ratios to obtain the corresponding measurements in inches. Body weight may be estimated by the chest girth formulas given on page 38.

Appaloosa

A special interest attaches to this strikingly colored spotted horse on account of its recent rise in popularity contrasted with its ancient origin. Paintings of spotted horses that resemble the present-day Appaloosa appear on the walls of a number of the caverns in France and Spain that were used by Cro-Magnon man as early as 18,000 B.C. The various types of wild horses that roamed the vast continent of Eurasia in those early Stone Age times were hunted by contemporary man for food, and for that reason were depicted, along with numerous other animals—bison, wild oxen, deer, even rhinoceros—on the walls of the caves with the belief that the hunting of them would thereby be successful.

During historic times, representations of spotted horses date back in Europe and China to at least 1000 B.C., and some of the paintings, carvings, and sculptures show coat patterns (spottings) of the distinctive type characteristic of the Appaloosa.

The name *Appaloosa* itself did not appear until Indians of the Nez Perce tribe, who lived in the regions of the Palouse River in Idaho, Washington, and Oregon, started breeding spotted horses from a few such animals that had been brought to the area by Spaniards about the year 1730. Thus, while the spotted horse is an ancient *breed,* the name Appaloosa—as derived from "a Palouse" —dates back only some 250 years.

It should be noted, however, that not all spotted horses are Appaloosas, and in order for an animal to be registered in the Appaloosa Horse Club, it must present various required characteristics, especially the types of coloration or spotting peculiar to the breed.*

In size and body build, the typical Appaloosa, as shown in Table 14, is seen to be a little lower at the shoulder (withers) and somewhat heavier than the Standardbred. Its build (which averages 0.935) is very similar to that of the Morgan (0.927), of which with the exception of its coloration, it is essentially a larger counterpart. And like the Morgan, the Arab, and the Quarter Horse, the Appaloosa is an efficient and versatile performer in all Western ranch and rodeo activities. It is also a popular pleasure riding horse and a fast racer under Quarter Horse conditions.

---

* The coat pattern in individual horses, however, is subject to great variation, and may range from almost complete spotting (head, neck, body, and limbs) to none at all. Usually the hoofs are vertically striped (black and white) and the sclera of the eye is white.

*Fig. 10. A Nez Perce Indian girl riding an Appaloosa horse, in a re-
enacted scene of a hundred years ago. Note the simple
yet decorative bridle and reins. The saddle is replaced by
a piece of hide, since the Nez Perce Indians did not pos-
sess saddles.* Photo by Johnny Johnston, courtesy the Appaloosa Horse Club.

Draft

It has long been the custom for writers on natural history to
differentiate light-built and heavy-built horses by the terms *warmblood* and *coldblood*, respectively. This differentiation has been applied not only to the numerous breeds of the domestic horse, but
also to the skeletal remains of various prehistoric or fossil forms of

TABLE 14
Measurements and Proportions of the typical Appaloosa*
(% = Percent of Withers Height)

	Measurement	Adult Males		Adult Females	
		in.	%	in.	%
A	Height, withers	62.00	*100.00*	61.25	*100.00*
B	Height, croup	61.75	*99.60*	61.07	*99.70*
C	Length, trunk (slantwise)	63.80	*102.90*	63.45	*103.60*
D	Length, head	24.49	*39.50*	23.94	*39.09*
E′	Width, forehead (max.)	9.37	*15.11*	9.16	*14.95*
F	Length, ear (externally)	7.51	*12.11*	7.35	*12.00*
G	Length, neck	29.44	*47.48*	28.97	*47.30*
H	Girth, neck (at throat)	31.73	*51.18*	28.95	*47.27*
J	Girth, chest (max.) (behind withers)	72.38	*116.74*	71.76	*117.16*
K′	Width, shoulders (max.)	16.43	*26.50*	15.91	*25.97*
L	Height, chest from ground	33.37	*53.82*	32.85	*53.63*
M	Height, elbow	36.68	*59.16*	36.42	*59.46*
N	Height, knee (top of pisiform)	19.21	*30.89*	18.92	*30.92*
P	Girth, forearm	19.79	*31.92*	19.01	*31.04*
Q	Length, back	29.28	*47.23*	28.97	*47.30*
R^P	Length, croup (projected)	20.84	*33.61*	20.66	*33.73*
S′	Width, croup (max.)	21.41	*34.53*	21.79	*35.58*
T^P	Length, hip to hock (projected)	36.93	*59.56*	36.65	*59.83*
U	Girth, gaskin (max.)	17.56	*28.34*	16.98	*27.72*
V	Girth, fore cannon (min.)	7.83	*12.63*	7.38	*12.05*
W	Girth, hind cannon (min.)	8.49	*13.69*	8.21	*13.40*
X	Height, hock	24.52	*39.55*	24.24	*39.58*
Y	Length, fore hoof (sole)	6.14	*9.90*	5.64	*9.20*
Y′	Length, hind hoof (sole)	6.00	*9.68*	5.52	*9.01*
Z	Width, fore hoof	4.99	*8.05*	4.59	*7.50*
Z′	Width, hind hoof	4.50	*7.26*	4.17	*6.80*
	Body Weight, lbs.	1150		1090	
	Body Build	0.946		0.924	

* For withers heights other than the average figures given in this table, use the percentage
ratios to obtain the corresponding measurements in inches. Body weight may be estimated by the chest girth formulas given on page 38.

Equidae. In many quarters, unfortunately, the need for recognizing certain natural differences between light or Oriental horses and heavy or Occidental ones has been replaced with the feeling that even a drop of "cold blood" in a light horse's constitution is bound to make the animal an inferior example of its breed—hence the widespread and often excessive emphasis put on the importance of "pure blood," and the discrimination levelled against any otherwise "warmblood" horse that is suspect as regards ancestry, irrespective of that horse's appearance and capabilities.*

But all this is an involved subject that cannot be gone into here. For our purpose, heavy (draft) horses can be differentiated from light (saddle or carriage) horses by certain criteria relating to general size, weight, and body proportions. To bring out some of these physical differences, let us compare the measurements and proportions of an Arab horse (Table 9) with those of a typically formed draft horse (Table 15). The latter measurements apply to either the Percheron, Belgian, Pinzgauer (Austrian), or Rheinish-Dutch breeds. Other draft breeds may vary slightly from this general physical type, but the degree of such variation is not usually significant.

In comparing the *percentage ratios*—not the absolute measurements—in Tables 9 and 15, it is found that one of the points of greatest differentiation is in the girth of the cannons in relation to withers height. Although there is no point at which the absolute cannon measurement in light horses ends and that in draft horses begins, the *ratio* of this measurement to the withers height is generally a sufficient basis by which to separate the two equine types. As will be seen, the relative fore cannon girths in stallions are: Arab *12.33*, Appaloosa *12.63*, Standardbred *12.66*, Thoroughbred *12.70*, Morgan *13.00*, Quarter Horse *13.24*, Shetland pony *13.53*, and Draft *15.23*. In one of the heaviest-built "warmblood"

---

* In this connection Dr. Dewey G. Steele states (p.20) : "The influence of a sire is approximately double that of a grandsire." And (p. 25) : "The current use of lengthy pedigrees lacks genetic significance. For all practical purposes individuals beyond the third gneration may be ignored." (*A Genetic Analysis of Recent Thoroughbreds, Standardbreds, and American Saddle Horses.* Bull. 462, Kentucky Agric. Exper. Sta., May 1944.) See also, by the same author: *The Influence of Ancestors.* Texas Livestock Journal, May 15, 1949.

However, as early as 1903, the English biometrician Karl Pearson had demonstrated with actual statistical data that the influence of successively earlier generations is not the generally-assumed .5000, .2500, .1250, .0625, etc., but is more nearly .6244, .1988, .0630, .0202. These ratios, which apply both to man and the horse, signify that the influence of the grandparents is .1988/.6244, or *less than a third* of the parents, rather than one-half, as was commonly thought. In view of this, the presence of a "cart horse" among ancestors only four generations removed should cause no concern as regards the heredity of a racehorse or other valuable equine. See *Law of Ancestral Heredity*, Biometrika, 2 (1903) : 296.

Fig. 11. Percheron stallion Orange Champion, owned by the Cafeteria Ranch, California. Note the typical draft proportions: thick neck and limbs, deep chest, long trunk, heavy quarters, and large hoofs. Courtesy Bone, Los Angeles.

horses—the German breed known as Oldenburg—the ratio is 9.02 in./64.45 in., or *13.99*. From these figures it is seen that a whole one percent (which amounts to about 0.80 of an inch) separates the cannon girth of a typical draft stallion from that of the possibly heaviest-built "warmblood" horse. Between the mares also of these light and heavy breeds a comparable difference in relative cannon girth prevails.

But an even more differentiating index than fore cannon girth/withers height is fore cannon girth/height of chest from ground.

TABLE 15
Measurements and Proportions of the typical Draft Horse*
(% = Percent of Withers Height)

	Measurement	Adult Males		Adult Females		Newborn Males		Newborn Females	
		in.	%	in.	%	in.	%	in.	%
A	Height, withers	64.85	100.00	64.37	100.00	40.43	100.00	40.13	100.00
B	Height, croup	65.37	100.80	64.81	100.68	41.39	102.38	41.27	102.83
C	Length, trunk (slantwise)	70.88	109.30	70.62	109.71	30.43	75.27	30.25	75.38
D	Length, head	27.50	42.40	27.33	42.46	14.73	36.43	14.68	36.58
E'	Width, forehead (max.)	10.37	16.00	10.30	16.00	5.83	14.42	5.79	14.43
F	Length, ear (externally)	7.98	12.30	7.92	12.31	6.02	14.90	6.00	14.93
G	Length, neck	30.15	46.50	29.97	46.56	13.34	33.00	13.30	33.14
H	Girth, neck (at throat)	36.59	56.42	33.48	52.01	19.29	47.70	19.10	47.59
J	Girth, chest (max.) (behind withers)	80.80	124.60	79.53	123.50	35.42	87.61	35.08	87.42
K'	Width, shoulders (max.)	20.70	31.92	19.14	29.73	9.39	23.22	9.31	23.19
L	Height, chest from ground	33.61	51.83	33.47	52.00	26.93	66.60	26.73	66.60
M	Height, elbow	37.09	57.20	36.83	57.21	27.90	69.00	27.77	69.20
N	Height, knee (top of pisiform)	19.91	30.70	19.76	30.70	15.82	39.13	15.77	39.30
P	Girth, forearm	21.69	33.45	20.79	32.30	13.69	33.86	13.50	33.65
Q	Length, back	32.60	50.28	32.58	50.61	14.17	35.05	14.13	35.21
R^P	Length, croup (projected)	23.08	35.59	23.00	35.73	9.47	23.42	9.42	23.47
S'	Width, croup (max.)	24.73	38.13	24.84	38.59	9.48	23.45	9.59	23.89
T^P	Length, hip to hock (projected)	40.66	62.70	40.34	62.67	22.02	54.46	22.02	54.87
U	Girth, gaskin (max.)	20.15	31.08	19.41	30.16	11.69	28.91	11.49	28.63
V	Girth, fore cannon (min.)	9.88	15.23	9.16	14.24	5.80	14.34	5.66	14.10
W	Girth, hind cannon (min.)	11.23	17.32	10.46	16.25	6.17	15.25	6.02	14.99
X	Height, hock	25.42	39.20	25.23	39.20	20.22	50.00	20.07	50.00
Y	Length, fore hoof (sole)	8.04	12.40	7.66	11.90	3.53	8.73	3.45	8.60
Y'	Length, hind hoof (sole)	8.04	12.40	7.66	11.90	3.60	8.90	3.52	8.77
Z	Width, fore hoof	6.48	10.00	6.18	9.60	2.87	7.10	2.80	6.99
Z'	Width, hind hoof	5.84	9.00	5.54	8.60	2.87	7.10	2.80	6.99
Body Weight, lbs.		1600		1484		145		140	
Body Build		1.150		1,083		0.430		0.422	

* For withers heights other than the average figures given in this table, use the percentage ratios to obtain the corresponding measurements in inches. Body weight may be estimated by the chest girth formulas given on page 38.

The latter measurement may conveniently be referred to as "leg length." This index yields these ratios in stallions: Arab *22.50*, Standardbred *22.89*, Thoroughbred *23.14*, Appaloosa *23.46*, Morgan *24.16*, Quarter Horse *24.75*, Shetland *26.32*, and Draft *29.40*.* The Oldenburg (see above), which is not included here with the breeds popular in the United States, has an index (as derived from the figures of von Nathusius) of 9.02 in./33.90 in., or 26.60. In individual draft horses bred for impressive exhibition size, the ratio may be as high as 34.00 or more. Thus it would appear that, barring occasional individual horses of odd conformation, most light horses have a Cannon girth/Leg length index of from 22 to 25, while in horses of the heavy draft type the index is usually somewhere between 28 and 31. In some of the heavier-built "warmblood" breeds, such as the Oldenburg, the index may be over 26 (in the Holstein it is 25.32), but is still appreciably below the level typical of draft breeds. It is interesting to note that in the Shetland it is over 26, which accords with the popular conception of the Shetland being a miniature draft horse. More specifically, in its general physical conformation the Shetland appears to be about 70 percent "warmblood" and 30 percent "coldblood" (see Shetland, following).

In the relative size of its hoofs also, the draft horse departs almost as far from light breeds as it does in the relative girths of its fore and hind cannons. This is probably the result both of its greater weight and the developing stress put on its hoofs from hauling heavy loads and needing maximum traction. In contrast, the foot bones of fossil equines of heavy build presuppose much smaller hoofs than those of existing (i.e., domestic) draft horses.

In general physical appearance and bodily proportions, a typical draft horse differs from an Arab or other light-built type in (1) its relatively long trunk (109–110 percent of its withers height, in comparison with 99–107 percent in light breeds), (2) a correspondingly long back (50–52 percent, in contrast to 45–47 percent), (3) long head (42–43 percent vs. 39–40 percent), (4) thick neck (56–57 percent vs. 49–54 percent), (5) great length from hip to hock (62–63 percent vs. 58–61 percent), and (6) relatively shorter limb bones below the level of the chest, and relatively longer ones in the trunk

---

* Indeed, the cannon girth/leg length index is practically as representative an index of general body build as that which we have used to determine the ratios given in the bottom row in Tables 9 to 16 inclusive. The formulas for the latter index are:

Body Build = Weight, lbs./Height, in. ^{3} x 196.1 (male) ; or x 194.7 (female) .

(i.e., humerus, scapula, femur).* Evident also is the great weight of the draft type, which rarely goes under 1400 pounds and is often over 2000, while in most light-built breeds the weight seldom exceeds 1400 pounds and averages only about 1000 pounds or so. And owing to the need for greater weight support in the draft horse, all the bones of its limbs, including the scapula and the pelvis, are inclined at more *upright* angles.

As the differences mentioned in the latter paragraph do not, for the most part, apply to the Shetland pony, the characteristics typical of that "miniature draft horse" are described in detail under the section "Shetland Pony" to follow.

As to the present status of the draft horse in the United States, it is gratifying to note that both in numbers and popularity it is on the increase. Breeding standards have become higher, with accompanying rises in price levels. The big horses are being used not only on farms, lumber camps, and other jobs where mechanized power is lacking, but also in such popular exhibition events as parades, the show ring, and load-pulling and plowing contests. Perhaps the best indication of the draft horse's "comeback" from once-threatened oblivion is that all five of the leading draft breed associations in this country—Belgian, Percheron, Clydesdale, Shire, and Suffolk—report increasing numbers of registrations.

Shetland Pony

In my comments on the draft horse, I mentioned that the Shetland is commonly regarded as a miniature likeness of that giant breed. So far as its capacity for hard work is concerned, the pony from the Shetland Islands is indeed a counterpart of the draft horse. This similarity applies particularly to the purebred Shetland used in the British Isles for work in coal mines (formerly), on the farm, and wherever else a small horse either for riding or drawing a cart is useful. The data in Table 16 are based mainly on Shetlands of that type, rather than the lighter-built, varicolored, crossbred ponies known as the American type. Of the latter type of pony, I have no body measurement statistics. That even the same Shetland can vary greatly in appearance, depending upon whether it is in its shaggy, winter coat or is groomed, is shown in Fig. 12.

---

* It should be mentioned, however, that some of the measurements in the Draft horse are relatively large only because they are here compared with the withers height rather than the trunk length. This applies particularly to the lengths of the head, neck, back, and croup, which are normally in close proportion to trunk length.

As to the origin of the true Shetland pony, the concensus of opinion is that it was *not* introduced into the islands by the Norsemen (c. A.D. 900), but was an indigenous animal known to the earlier Celtic inhabitants long before the Norwegian invasion. Earlier still, there is evidence that the Shetland may have descended from the Celtic pony of postglacial northwestern Europe. The small size of the Shetland is generally attributed to its long residence on small islands that are sparse in vegetation and subject to harsh weather. For example, until the Shetland ponies were domesticated and fed by man, their only food during the winter was whatever herbs they could eke out from under the snow, plus the kelp or seaweed that was washed ashore. As a parallel, during Pleistocene times on the Channel Islands off the coast of Southern California, dwarf elephants existed, presumably stunted in size for the same reason: insular isolation.

Physically, on a smaller scale, the Shetland is similar to the draft horse in its (1) large head,* (2) rather short neck, (3) long trunk and back, (4) long hip to hock distance, (5) low chest-from-the-ground height, (6) thick neck, and (7) large chest. In relative head length, forehead width, and ear length the Shetland is somewhat larger than the draft horse, but this is a result of the geometric law that requires a *relatively* larger brain, head, eyes, ears, etc. in the smaller representatives of a given type of mammal, including man. In various other features—such as leg length, shoulder width, croup width, and the girths of the forearm, gaskin, and cannons—the Shetland is intermediate between the draft type and the typical light-built horse.

In absolute size, the greatest divergence of the Shetland from the draft horse is in its *feet* (cannon bones, pasterns, and hoofs), which from a geometric standpoint could be said to represent those of a draft horse if this giant equine were reduced to one-fourth or one-fifth of its normal weight, this for the reason that a quadruped's feet are generally of a size proportionate to the *body weight* of the animal. This law of supporting strength in the limbs applies to land animals of all builds, from the slender antelopes and gazelles to the ponderous rhinoceroses and elephants. Thus it accounts for the relatively small feet of ponies in comparison with the huge, weight-bearing feet of heavy draft horses.

---

* The head is especially large in the newborn foal, with a relative length of 45–46 percent as compared with 40 percent or less in most other breeds. This heavy head in turn requires an exceptionally thick neck (c.48 percent) for its support.

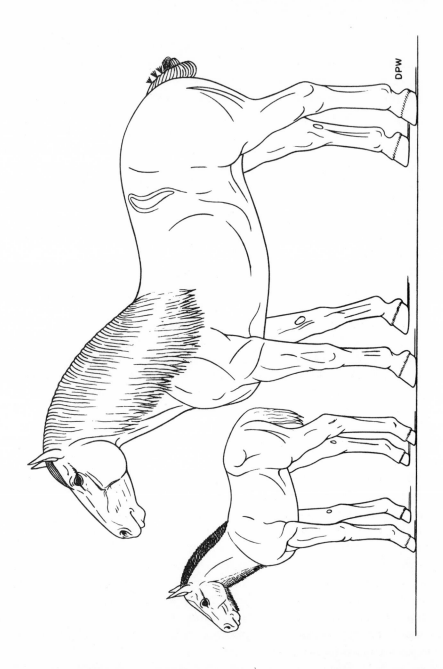

Fig. 12. Relative size and body proportions of a draft (Percheron)
mare and foal. Compare with Figs 5 and 14. x .06 nat-
ural size.

During recent years, there has been increased activity in some quarters in the breeding of ponies of the smallest possible size. Since most of these very small ponies are Shetlands there is consequently a great variation within this breed between the largest and the smallest specimens. The typical shoulder (withers) height may be said to be from 38 to 42 inches. A purebred Shetland, to be registered, must not exceed the latter height. Unfortunately, in the United States there appears to be a large proportion of crossbred ponies of heights up to 46 inches (and with an unduly high percentage of piebald and skewbald individuals) which are nevertheless referred to as genuine Shetlands. So much does this condition prevail that to many a person the term "pony" is synonymous with Shetland pony. While parti-colors may indeed occur in purebred Shetlands, the basic color is *black,** with variations of bay, brown, sorrel, and dappled grey. In Germany, about one Shetland out of every seven is *white*.

The measurements of typical Shetlands given in Table 16 have been derived from the body *proportions* shown by the data of J. E. Flade (1959), applied to the *average* heights of purebred stallions, mares, and newborns, this for the reason that Flade's figures indicate (on German Shetlands) ponies of slightly larger adult heights and weights than those typical of Shetlands bred in the British Isles. The measurements given by Flade for living ponies were checked by the writer against measurements of 7 skulls and 3 to 5 postcranial skeletons recorded by other hippologists.

In artificially inseminated crosses between Shetland ponies and Shire (English draft) horses, the foals of Shire mares were found to weigh about the same as purebred Shire foals. That is, the physique of the mother appeared largely to control the size of the offspring. On the other hand, the foals born to Shetland mares were only about one-third the weight of the Shire-mothered foals. Here again was shown the predominating influence of the female parent. The Shetland-mothered foals grew slower than purebred Shire foals, but nevertheless much faster than purebred Shetland foals.

A "midget" pony is one under 32 inches in height. Mr. Smith Mc-Coy, of Roderfield, West Virginia, has (or had, in 1965) a herd of forty midget ponies, among which was a stallion of 28½ inches and a mare, named Sugar Dumpling, of only 20 inches and 30 pounds, making it probably the smallest adult horse in the world in its day.

---

* The so-called black coloration common to purebred Shetlands is not jet black, but what is known locally as Shetland black, a somewhat lighter, more slaty hue.

Sugar Dumpling died sometime in 1965. More recently, still smaller ponies have been given publicity, but most of such animals have been foals—sometimes, very young ones—rather than full-grown, adult specimens.*

The derivation of the Shetland from the Celtic pony of earlier times is indicated in one respect by the usual absence in both types of the chestnuts on the hind legs. In individual ponies in which these hind chestnuts are present they are very small. Dr. J. C. Ewart had a Shetland mare in which the right hind chestnut measured only .40 x .16 of an inch, while the left was only .20 inch in diameter. The ergots (fetlock callosities) on the forelegs were absent, while those on the hind legs were very small.

The Tennessee Walking Horse and the American Saddle Horse

These two popular American breeds are of such average or typical height and weight for saddle horses that there is no need to list their various body measurements in separate tables. The Tennessee Walking Horse (which is in some quarters called Plantation Walking Horse) stands on the average (in stallions) about 62 inches at the withers and weighs about 1100 pounds. This height and weight would presuppose a chest girth of about 71.5 inches, forearm 19.5 inches, gaskin 17.25 inches, and fore cannon 7.85 inches. The various height measurements—croup, elbow, chest from ground, hock—in the Walking Horse should be about the same as those in the Appaloosa (Table 14). At 62 inches and 1100 pounds, the body build in Walking Horse stallions is 0.905, which is about midway between that of the Standardbred and the Morgan.

In the more slender-built American Saddle Horse, which averages 62 inches and 1000 pounds, the heights, like those in the Walking Horse, are about the same as those in the Appaloosa. The corresponding girth measurements in Saddle Horse stallions would be: chest 69 inches, forearm 18.9 inches, gaskin 16.4 inches,

---

* Breeders of miniature ponies commonly state also that these equines are "mature" at two or three years of age. In human midgets, it has been found that completeness of growth and development lags *behind* that of normal-sized individuals. In view of this, it would appear strange if in midget ponies or horses this morphological pattern were reversed. One might say that three-year-old racehorses are "almost" mature, yet by five years they would normally be an inch taller and about a hundred pounds heavier. In a 30-inch-tall miniature pony, the proportionate gains (after three years) would be only about a half-inch and ten pounds. The latter increases apparently are sufficiently small to be ignored by most midget pony dealers.

TABLE 16
Measurements and Proportions of the typical Shetland Pony*
(% = Percent of Withers Height)

	Measurement	Adult Males		Adult Females		Newborn Males		Newborn Females	
		in.	%	in.	%	in.	%	in.	%
A	Height, withers	39.84	100.00	39.37	100.00	26.10	100.00	25.96	100.00
B	Height, croup	40.04	100.50	39.74	100.94	26.49	101.50	26.36	101.55
C	Length, trunk (slantwise)	43.23	108.50	43.23	109.80	19.84	76.00	19.77	76.15
D	Length, head	17.65	44.30	17.41	44.02	11.92	45.67	11.81	45.50
E'	Width, forehead (max.)	6.85	17.20	6.76	17.17	4.44	17.01	4.40	16.95
F	Length, ear (externally)	5.30	13.30	5.23	13.28	4.50	17.24	4.46	17.18
G	Length, neck	18.37	46.11	18.19	46.20	8.66	33.20	8.63	33.24
H	Girth, neck (at throat)	23.25	58.36	21.58	58.41	12.57	48.16	12.44	47.92
J	Girth, chest (max.) (behind withers)	49.64	124.60	49.64	126.10	22.76	87.20	23.08	88.90
K'	Width, shoulders (max.)	12.61	31.65	11.89	30.20	5.61	21.50	5.61	21.60
L	Height, chest from ground	20.48	51.41	20.01	50.83	17.40	66.67	17.14	66.03
M	Height, elbow	23.38	58.68	22.90	58.18	18.01	69.00	17.80	68.57
N	Height, knee (top of pisiform)	11.99	30.10	11.64	29.56	10.05	38.50	10.00	38.52
P	Girth, forearm	12.80	32.13	12.22	31.04	8.57	32.84	8.47	32.64
Q	Length, back	20.16	50.60	20.16	51.21	9.19	35.20	9.18	35.36
R'	Length, croup (projected)	12.63	31.70	12.63	32.08	6.14	23.53	6.13	23.60
S'	Width, croup (max.)	14.34	35.99	14.47	36.75	6.08	23.28	6.24	24.04
T'	Length, hip to hock (projected)	24.74	62.10	24.62	62.53	13.95	53.45	13.89	53.51
U	Girth, gaskin (max.)	11.66	29.27	11.26	28.60	7.45	28.54	7.32	28.20
V	Girth, fore cannon (min.)	5.39	13.53	5.19	13.17	3.85	14.77	3.80	14.64
W	Girth, hind cannon (min.)	5.96	14.96	5.70	14.48	4.18	16.01	4.12	15.87
X	Height, hock	15.78	39.61	15.60	39.62	13.05	50.00	12.98	50.00
Y	Length, fore hoof (sole)	4.50	11.30	4.17	10.60	2.09	8.00	2.05	7.90
Y'	Length, hind hoof (sole)	4.54	11.40	4.25	10.80	2.19	8.40	2.15	8.28
Z	Width, fore hoof	3.63	9.10	3.39	8.60	1.70	6.52	1.67	6.45
Z'	Width, hind hoof	3.27	8.20	3.03	7.70	1.70	6.52	1.67	6.45
Body Weight, lbs.		371		361		45		44	
Body Build		1.150		1.152		0.496		0.490	

* For withers heights other than the average figures given in this table, use the percentage ratios to obtain the corresponding measurements in inches. Body weight may be estimated by the chest girth formulas given on page 38.

Fig. 13. (Above) Shetland ponies in a semiwild state in the Scottish Highlands. (Below) The same two ponies after having been trained and groomed. Photo from J. W. Axe. The Horse, Its Treatment in Health and Disease. London: 1906.

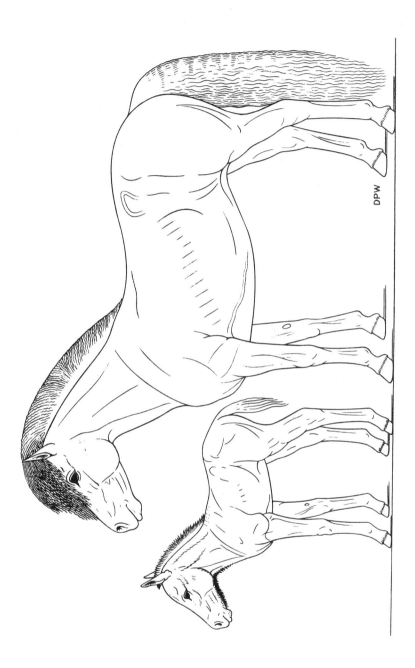

Fig. 14. Relative size and body proportions of a purebred Shetland mare and foal. Compare with Figs. 5 and 41. x .06 natural size.

and fore cannon 7.5 inches. The body build is only 0.823, which means that the American Saddle Horse is slightly more slender than the Arab.

However, the measurements given in Table 17, which are the *averages* of the six different light-horse breeds named, could well represent those of either a Tennessee Walking Horse or an American Saddle Horse, each of which closely approaches this average of six breeds. The elegance of appearance in these show horses, along with the perfection of their gaits, distinguish them more than does their "average" bodily dimensions.

5
Sex Differences of Size and Bodily Proportions in Adult Horses of Light Breeds

TYPICAL SEX DIFFERENCES IN THE BODY MEASUREMENTS OF ADULT horses are listed in Table 17. The measurements presented are the average of six light breeds: Arab, Thoroughbred, Standardbred, Quarter Horse, Morgan, and Appaloosa. In determining these averages, the draft horse and the Shetland pony have been omitted, since the latter two breeds are equine *types distinct in size* from the usual type of saddle and carriage horses. However, in both draft and Shetland the sex differences correspond closely with those indicated in Table 17. If, on the contrary, the Shetland were added to the above-mentioned six breeds of light horses, the average withers height in males would decrease from 61.30 inches to 58.23 inches, the latter height being below even that of the Arab, the smallest of typical light horse breeds. Again, if the Draft horse were added to the aforementioned six light horses, the average weight would increase from 1090 pounds to 1163 pounds, the latter weight being that of a Thoroughbred—one of the heaviest of light horse breeds —rather than a figure that represents the average or typical weight of such breeds.

Going down the list of measurements in Table 17, it is seen that in most respects the adult female horse is from 96 to 99 percent as large as the typical stallion or gelding (of the same breed). In chest girth the female is generally only slightly smaller than the male, although in individual horses there is a considerable amount of variation in this ratio. Frequently, in Arabs and Anglo-Arabs, as well as several other light breeds, the chest girth of mares is greater than that of stallions. Only in croup width is the female consistently larg-

Table 17
Comparison of the Bodily Measurements of Adult Male and Female Horses
of Light Breeds
as *Averaged* from Tables 9, 10, 11, 12, 13, and 14, inclusive
(% = Percent of Withers Height)

	Measurement	Adult Males		Adult Females		♀ meas., in. ÷ ♂ meas., in. in %
		in.	%	in.	%	
A	Height, withers	61.30	100.00	60.54	100.00	98.78
B	Height, croup	61.04	99.58	60.50	99.93	99.12
C	Length, trunk (slantwise)	62.58	102.10	62.25	102.82	99.47
D	Length, head	24.10	39.32	23.68	39.11	98.26
E'	Width, forehead (max.)	9.20	15.01	9.02	14.90	98.04
F	Length, ear (externally)	7.36	12.01	7.23	11.95	98.23
G	Length, neck	29.27	47.78	28.93	47.79	98.82
H	Girth, neck (at throat)	31.22	50.93	28.87	47.69	92.47
J	Girth, chest (max.) (behind withers)	71.09	115.97	70.55	116.53	99.24
K'	Width, shoulders (max.)	16.37	26.75	15.77	26.05	96.33
L	Height, chest from ground	33.28	54.31	32.74	54.08	98.38
M	Height, elbow	36.49	59.54	36.23	59.84	99.29
N	Height, knee (top of pisiform)	19.09	31.14	18.81	31.08	98.56
P	Girth, forearm	19.45	31.73	18.68	30.85	96.04
Q	Length, back	28.57	46.61	28.33	46.80	99.18
R^P	Length, croup (projected)	20.48	33.41	20.30	33.53	99.12
S'	Width, croup (max.)	21.02	34.30	21.43	35.40	101.93
T^P	Length, hip to hock (projected)	36.46	59.49	36.18	59.76	99.23
U	Girth, gaskin (max.)	17.19	28.04	16.64	27.49	96.79
V	Girth, fore cannon (min.)	7.81	12.74	7.40	12.22	94.75
W	Girth, hind cannon (min.)	8.48	13.83	8.22	13.57	96.93
X	Height, hock	24.34	39.71	24.04	39.69	98.75
Y	Length, fore hoof (sole)	6.08	9.92	5.56	9.18	91.53
Y'	Length, hind hoof (sole)	5.93	9.68	5.44	8.99	91.63
Z	Width, fore hoof	4.91	8.01	4.54	7.60	92.44
Z'	Width, hind hoof	4.40	7.19	4.10	6.77	93.08
	Body Weight, lbs.	1090		1033		94.77
	Body Build	0.928		0.907		97.96

er than the male.* This holds true even in Draft breeds, where in some measurements, such as withers height, there is less difference between the sexes than in the light horse type. The female superiority in the single respect of croup width is present even in newborn foals. The greatest sex difference occurs in the girth of the fore cannon, where the female averages less than 95 percent of the male, and in the width and length of the hoofs, where a minimum ratio of 93 to 91 percent is typical. In weight also, the mare averages less than 95 percent of the male, although in many instances she is the heavier of the two, which during pregnancy is only to be expected. In draft breeds the weight ratio is typically only about 93 percent, while in the Shetland pony it may be from 97 to 100 percent or over. This would indicate that the greater the absolute size of the breed or type, in many respects the greater the superiority in size of the male, and vice versa.

In most of the measurements of *height* or *length*, there is marked uniformity among the numerous light breeds. For instance, in croup height females generally show a higher ratio to withers height than males, simply because in females the withers (i.e., the fourth to sixth spines of the dorsal vertebrae) are less developed than in males. This in turn is probably the effect of having to support a less-heavy head and neck. In elbow height the average ratio to withers height is about 60 percent in both sexes, although in the draft and the Shetland the ratio is somewhat less on account of the vertical depth of the chest being greater. That is, the height from the ground of the top of the elbow is generally five percent or so greater than the height to the bottom of the chest, so that in relatively large-chested breeds or types (such as the draft and the Shetland) elbow height in relation to withers height is less, and in smaller-chested or "speed" types relatively greater, in conformity with longer legs.

Hock height is even more consistent than elbow height, being in light breeds of both sexes about 40 percent, and in draft breeds about 39 percent, with Shetland ponies showing an intermediate ratio of 39.6 percent. In the proportions of the cannons and hoofs, the size is essentially in ratio to the body weight, with female horses showing consistently smaller ratios than males. This also is a sex difference that prevails from birth on. But why, for a given body weight, the hoofs of mares should be smaller than those of stallions or geldings, remains to be explained.

---

* Because of this, one of the most differential sex ratios in horses is croup width/neck girth. This ratio, in both light and draft breeds, averages about 10 percent greater in females than in males.

6

Body Measurements and Proportions of the Newborn Foal as Compared with the Adult

FIGURE 5 SHOWS THE RELATIVE SIZE AND BUILD OF AN ARAB MARE AND newborn foal, both figures being based on the measurements listed in Table 9 (cf. also Fig. 14). Tables 10 (Thoroughbred), 12 (Quarter Horse), 15 (draft), and 16 (Shetland) likewise include measurements and proportions of newborn foals. These foals, although of differing breeds, have such similar physical characteristics that they all may be described as follows.

To start with the head of the foal: (1) the hinder portion or cranium is relatively longer and the facial portion relatively shorter than in the adult horse; (2) the eyes of the young animal are thus more forwardly placed, as well as being relatively larger; (3) the forehead is more bulging; (4) the ears are relatively much larger, while (5) the mouth and lips, not yet used for grazing, are noticeably small. The mane and the tail are short and scanty. Often, however, there is long, fuzzy hair over the body, especially on the jaws and neck. A newborn foal has only 16 teeth as compared with the normal 40 of its dam and 44 of its sire.

But it is in the *limbs* of the young foal that the greatest differentiation from the adult is most evident. Whereas elbow height in the adult is generally about 60 percent of the withers height in light horses and 57 or 58 percent in heavier breeds, in the newborn foal of either light or draft breed the elbow height is no less than 69 percent. This means that the elbow height in the newborn animal is over 74 percent of that in the adult. In the hind limbs an even greater proportionate length prevails in hock height, this measurement in the adult averaging about 40 percent of the withers height and in

90

Fig. 15. A Tennessee Walking Horse and rider. Painting by John Mariani, courtesy Mrs. John Mariani.

the newborn no less than 50 percent. This brings up the commonly believed idea that the legs of a newborn foal are as long as they will ever be. But as is seen in Figure 3, a small amount of growth in the length of the metacarpal (fore cannon) bones continues to take place until about a year of age. This growth, in an Arab foal, amounts to about 1¼ inches, and in foals of larger breeds, such as the Thoroughbred, to slightly more (see also p. 49).

The relatively great length of the legs in a newborn foal makes the length of its entire spinal column (neck, back, and croup) correspondingly short in relation to its withers height. For while the trunk length in adult light horses is generally at least equal to the withers height, in newborns of both light and draft types the relative trunk length is only about 76 percent. Another noticeable dif-

ference in the foal is the relatively small hoofs. The body **weight at**
birth, as explained on page 44, varies in proportion to the size of the
parents, the offspring of ponies or small horses being *relatively* heavier than those of draft breeds.

Fig. 16. An American Saddle Horse and rider. Painting by John Mariani, courtesy Mrs. John Mariani.

7

Body Measurements and Other Data on Some Noteworthy Horses of Various Breeds

WHILE THE AVERAGE BODY MEASUREMENTS DERIVED FROM A SERIES of horses of a given breed best express the bodily conformation typical of that breed, it is also desirable to know the specific measurements of outstanding individual performers, or sires, within the various breeds, so that certain advantageous characteristics may be striven for in the offspring of such animals. In racehorses, for instance, it would appear that the best performers stand somewhat taller than the average, Man o' War and Secretariat being examples of this among Thoroughbreds and Greyhound among Standardbreds. This advantage in height and length of leg holds true among human sprinters also, since evidently the slightly slower leg movements in such taller runners are more than compensated for by their longer strides.

In the Arabian horse, on the other hand, increased body size commonly results in coarseness of appearance. Accordingly, for elegance and vivacity the typical height of 58–60 inches and weight of 860–940 pounds is still the ideal in this historic breed.

In the draft horse, heights and weights substantially above the average are common in individuals bred for exhibition purposes. However, the strongest drafters, as determined by pulling contests, need not weigh over 2000 pounds, while for an all-around farm horse a weight between 1500 and 1600 pounds has been proven the most efficient.

In ponies, a wide range in body size prevails among the various breeds, and the size to be chosen depends upon the proposed use to

93

which the pony will be put. Mounts for small children may be correspondingly small, but should nevertheless be chosen with the prospect in mind that the child may *outgrow* the pony. At present in some quarters there is a demand for ponies of the smallest obtainable sizes, which may be such that the animals are even allowed to roam inside the house. But such miniature ponies, or horses, are decidedly a novelty, and usually a rather expensive one.

My purpose now is to cite a few well-known horses the measurements of which have come to my attention and which are not ordinarily or conveniently available. Too, I am including various other items of information not generally presented in books about horses. While some of this may have no direct bearing on either growth or nutrition, I hope that it will be of interest.

Arabs

Mesoud, a famous purebred Arabian sire, at the age of ten years stood 58 inches and weighed 990 pounds; his chest measured 69 inches. Much larger was Khaled, who stood 63.5 inches, had a 72.5-inch chest, and weighed 1160 pounds. Khaled's mounted skeleton is in the American Museum of Natural History, New York City. In the National Museum, Washington, D.C., is the skeleton of Haleb, who died at the age of eight years and whose bones indicate that he stood in life about 1490 mm (58.65 inches). The tallest well-known Arab stallion was perhaps Nureddin, who stood 63.75 inches, the height of an average-sized Thoroughbred. As to the racing speed of the Arab horse, perhaps the fastest galloper was Sartez, who in 1948, ridden by Buck Griffin, set thirteen new world records at distances ranging from 4½ furlongs to 2400 meters. On the basis of size, one would expect the speed of an Arab to be at least 96 percent of that of a Thoroughbred, yet the records made by Sartez average only about 88 percent, even taking into account that Sartez performed twenty-five years ago. This should be conclusive proof that the specialized training of Thoroughbreds over a period of nearly three hundred years has produced an equine breed of unequalled racing speed.

Thoroughbreds

Man o' War, one of the greatest Thoroughbreds, as a three-year-

Fig. 17. A three-day-old Arab colt. In this photo are brought out many details of a typical newborn foal's body structure and conformation. Courtesy the W. K. Kellogg Arabian Horse Ranch, Pomona, California.

old stood 65.6 inches, had a chest of 71.75 inches, and weighed 1150 pounds. At five years (1922) he had increased to 66.25 inches and 1370 pounds, with a 76.5-inch chest. The great recent-day champion, Secretariat, appeared to be, at three years, about the same size as was Man o' War at that age. Considerably smaller were Seabiscuit, who at three years (1938) stood 62 inches, had a 70-inch chest, and weighed 1040 pounds; and War Admiral, who though a son of Man o' War, was only 62.5 inches and 960 pounds (in 1937). Whirlaway at four years (1942) was 63 inches and 1070 pounds, with a chest of 71 inches. At the age of nine, and at stud, he had increased to 64 inches and 1185 pounds, with a chest of 73 inches. Another great champion, Equipoise, at four years (1932) stood 63.75 inches and had a 73-inch chest, which would indicate a weight of about 1180 pounds. Gallant Fox, winner of the Triple Crown in 1930, was 65 inches and 1125 pounds. One of the smallest top Thoroughbreds was Assault, who stood only about 60.5 inches; while the tallest Thoroughbred was Limbo, who towered 73 inches. One of the heaviest Thoroughbreds was probably Bull Lea, who as a three-year-old weighed 1250 pounds and later went to over 1400. The Thoroughbred appears to be one of the few breeds of horses that has significantly increased in height during the last century. In the Thoroughbred, perhaps by artificial selection for greater speed, the average withers height has increased from about 62 inches in 1900 to about 65 inches in the United States today (and perhaps since about 1940).

Standardbreds

Messenger, a grey Thoroughbred stallion imported in 1788, was the chief progenitor of what came to be a leading line of trotters. Messenger's height was 62.75 inches, which was well above the average for Thoroughbreds in his day. The greatest of recent trotting champions was Greyhound (1:55¼), who in 1936, at the age of four years, stood 65.25 inches, measured 70.5 inches in chest girth, and weighed 1100 pounds. Spencer Scott (1:57¼), a previous holder of the mile trotting record, in contrast to Greyhound stood only 63.25 inches yet was heavier built, with a chest of 71.75 inches and a weight of 1120 pounds (1941). Of the greatest pacer, Dan Patch (1:55¼ in 1905), I unfortunately have no measurements.

Quarter Horses

While Quarter Horses generally range in weight between 1050 and 1300 pounds, some outstanding individuals have ranged well outside these usual limits. Steeldust's Cowboy, for example, weighed only 925 pounds at a height of 56 inches, while the almost legendary Peter McCue in his later years scaled no less than 1430 pounds at a height of 66 inches. Despite his size, he was very fast. In Quarter Horse racing, the distance is generally not over a quarter-mile, on a straightaway track, and the horses go into action from a *standing* start. This is in contrast to Thoroughbred and Standardbred racing, in both of which there is a *running* start. The length of the preliminary run varies from track to track according to the distance of the race, and also by the need for the starting gate to get all the horses off to a fair start. Quarter Horses are usually faster than Thoroughbreds over short distances, but only because they have been trained to *start* more quickly. Once under way, there is little to choose between a championship-class Quarter Horse and a championship-class Thoroughbred.

Morgan Horses

The definitive study of this breed made by D. C. Linsley in 1857 showed that the typical height of stallions was 60 inches and the typical weight 1035 pounds. On the basis of these figures it is interesting to surmise what the founding sire of the breed, Justin Morgan (1789–1821), probably weighed at his asserted height of only 56 inches. His weight is usually stated—presumably on the authority of Linsley—to have been 950 pounds. However, this poundage would have made Justin Morgan (or Figure, as he was then called) almost as heavy-built as a draft horse, a proportion which certainly is not typical of the Morgan breed. According to Robert West Howard (*The Horse in America*, p. 123), "Later researchers estimate Figure could not have weighed much more than eight hundred pounds." If Figure had been of *typical* Morgan body-build, at his height of 56 inches he would have weighed $(56/60)^3$ x 1035, or 841 pounds. So, at that weight we must leave Figure, later Justin Morgan, the remarkable single progenitor of an entire equine breed.

Appaloosa Horses

According to registrations in 1968, the Appaloosa was the third most popular breed in the United States, being topped only by the Quarter Horse and the Thoroughbred. Like several other versatile breeds, the Appaloosa is being used the world over as a stock horse and for pleasure riding, all-around ranch work, trail riding, hunting, jumping, and racing. In the latter sport, which is conducted along

the same lines as Quarter Horse racing, the speed of Appaloosa horses over distances from 220 to 440 yards averages only about three
percent slower than that of Quarter Horses, which have been competing in racing contests for over three hundred years. To illustrate,
at 350 yards the Quarter Horse record (as of March, 1973) is 17.33
seconds and the Appaloosa record 17.90 seconds. It is interesting to
note that horses of such decidedly different physiques can, like human sprinters, be almost equally fast on the track; for the average-
sized Quarter Horse stallion stands only 60 inches high and weighs
1178 pounds, while the average Appaloosa stallion stands two inches taller yet weighs 28 pounds less. Then there is the still more slender Thoroughbred which also, when trained in making a fast start,
can cover short distances as rapidly as any Quarter Horse. It may
be that in such short races a long body (as in the Quarter Horse)
confers as much of an advantage as do long legs (as in Thoroughbreds).

Draft Horses

The size to which a draft horse may be bred, or will attain, depends largely upon the *use* to which the animal will be put. Today
there is only a limited demand for horses to perform the heavy work
of hauling, plowing, etc. for which they were formerly used on
farms and in industry. However, many draft horses are today being
bred for *competition* in pulling and plowing exhibitions, a great
number of which take place through the summer and fall months in
many states, especially Michigan, Iowa, Indiana, Ohio, and New
York. While the paired drafters in some pulling teams stand as tall
as 70–72 inches and each weigh 2200 pounds or more, in 1940 a pair
of geldings named Sam and Prince, each standing only 64 inches
and weighing together 4150 pounds, made a world record by pulling
4025 pounds the full distance of 27½ feet. By 1972, the record poundage pulled had been increased to 4375 pounds, but the combined
weight of the two horses that accomplished it was a hefty 4400
pounds. An even heavier team, weighing 4620 pounds, pulled 4100
pounds to win the National Heavyweight Contest for 1972. Greater
strength in proportion to body weight was, as would be expected,
shown by a light-weight team weighing 2990 pounds, which pulled
3900 pounds. On the basis of the muscular cross-section indicated
in the latter team, a pull of 4400 pounds *should* be possible
for a team weighing only 3590 pounds. Evidently, as in the strong-

Fig. 18. A Belgian mare and her twin (once in 10,000 births) colts. This photograph shows clearly the body proportions of a newborn draft foal. Courtesy the Peffer Farm Stables, Stockton, California.

est human athletes, extreme muscular size is generally accompanied by a large amount of subcutaneous fat, which of course increases the body weight and consequently reduces the ratio of strength to weight. Probably the heaviest draft horse was Brooklyn Supreme, a Belgian stallion that stood 78 inches and weighed 3200 pounds. Probably even taller, and nearly as heavy, was the mare Nebraska Queen, who scaled "more than 3100 pounds."

Shetland Ponies

Although formerly much used, especially in the British Isles, as a small-sized work horse, the Shetland today is in demand chiefly as a children's pony and in the show ring. The wide variation in both size, build, and coloration in this pony breed makes it impossible to define more than the average characteristics of the British-type Shetland, which we have done herein on page 81. While the maximum permissible height in the American-type Shetland is 46 inches, there is no lower limit, and some ponies, like the late mare Sugar Dumpling, have measured as little as 20 inches high at the shoulder. At birth, one Shetland stood only $17\frac{1}{4}$ inches and weighed a mere 12 pounds. There is a Miniature Horse Racing Association, and also the United States Pony Trotting Association. In the latter, ponies up to 48 inches in height may compete, and there are numerous classifications for this and lesser heights. The racing is done on either half-mile or quarter-mile tracks. An analysis of the best pony racing records reveals that the fastest *galloper* did a quarter-mile, on a straight course, in 26 seconds, carrying an 80-pound rider. If this pony had been only 41 inches tall, his time of 26 seconds would be exactly proportionate to that of a 60-inch-tall Quarter Horse doing a quarter-mile in the present World Record time of 21.46 seconds. However, the probability is that the pony was taller than 41 inches (his height not having been stated), possibly even 48 inches.

In *trotting*, the smaller-sized ponies show slower times than even their lesser heights would indicate. For example, the best half-mile trotting record for a pony under 40 inches in height (in this case $39\frac{1}{2}$ inches) is 1:57.6 (117.6 seconds), while that of a pony standing $47\frac{3}{4}$ inches is 1:31.4 (91.4 seconds). If the latter pony had trotted at the same rate as the $39\frac{1}{2}$-inch pony, in proportion to his effective size, his time would have been 39.5/47m75x 117.6, or 107 seconds rather than the 91.4 seconds he actually registered. This comparison shows the relatively superior speed of the

taller pony trotters, or conversely, the relative slowness of the smaller ponies. Possibly this denotes a genetic or racial influence, just as
occurs among human athletes, in which the highest ratio of strength
to body weight (hence, *speed*) is shown in performers of *medium
size* rather than in those either markedly smaller or markedly larger
than the average.

If we assume that Uhlan, the holder of the Standardbred trotting
record at one-half mile, of 56¼ seconds, had an average-length stride
of 21 feet and so took 2.23 strides per second and that Uhlan's
height was (for trotters in 1911) an average 62 inches, then a pony
39½ inches high should—on the basis of the law of the pendulum*
—take 2.79 strides per second and trot the half-mile in 70.6 seconds.
As we have seen that the half-mile trotting record for a pony 39½
inches high is (or was, in 1964) actually 117.6 seconds, it would appear that such a pony takes only 70.6/117.6, or 1.67 strides per second instead of the expected 2.79.** This is only sixty percent of the
speed that would be expected in a pony 39½ inches high on the basis of a Standardbred trotter 62 inches high. Thus it would seem that
small ponies, especially heavy-built ones of the purebred Shetland
type, are "miniature draft horses" in the matter of trotting speed as
well as bodily conformation.

Small ponies now engage even in pulling contests, and there is
a Central Kansas Shetland Pulling Association, in Newton. The object pulled is a sled loaded to equal the weight of the two ponies.
As the ponies move each load, more weight is added to the sled until it can no longer be pulled. In this manner a team of average-sized but strong Shetlands, weighing together 750 pounds, should
be able to pull 1500 pounds. Also, in various parts of the country,
Shetland owners stage chariot races, chuck wagon races, etc. with
their ponies just as are staged at Western rodeos with full-sized
horses.

---

* The speed of a pendulum is in ratio to the *square root* of its length; hence a pendulum
10 inches long would move twice as fast as one 40 inches long, and so on. The same
relationship applies to the rapidity of leg movements (re their length) in humans,
horses, greyhounds, and all other running animals.

** Either that, or takes about the estimated 2.79 strides per second, but with a stride only
60 percent as long as would be expected—that is, a stride of only about 8 feet instead of
over 13 feet.

8

Heights and Weights
of Horse and Pony Breeds
Common in the United States

ALTHOUGH THE WRITER HAS DERIVED MUCH OF HIS BASIC DATA ON
the measurements of adult horses from Tables 1 and 2, the horses
listed in those tables are mostly European breeds. It has therefore
appeared desirable to list also the typical body sizes of breeds popular in the United States, even though the measurements are confined to height and weight. It is rare indeed to see in books about
American horses any reference to measurements other than these
two basic ones.

Table 18 lists—so far as it has been possible to establish them—
the heights and weights ordinarily found in various breeds or types
of horses and ponies used in the United States today. While most of
these horses are represented by specific breed associations, it is nevertheless unusual to see in the descriptive literature any measurement other than height, which is invariably expressed in the conven-
tional measure of "hands" (a "hand" being four inches).* In view
of this paucity of authentic data, many of the measurements listed
in Table 18 are estimations based on what the writer has interpret-
ed as the most reliable sources of information. But the figures are at
least consistent with whatever is definitely known of the breeds.
Withers height in 100 cases ordinarily ranges about ± 5 percent; and
body weight, accordingly, about ± 16 percent. If writers about
horses would only bear in mind that weight varies as the *cube* of
height or length, and as either the cube or the *square* of girth (since
girth encompasses *cross-section*), there would be less confusion and
error in the statements some of them make concerning weight.

---

* Why horsemen in Great Britain and the United States continue to use such an out-
moded measure is difficult to understand.

TABLE 18
Typical Heights and Weights of the More Common Breeds
and/or types of Horses and Ponies in the United States
Listed according to body build.   Sexes averaged.

Breed or Type	Height, inches Average	Range*	Weight. lbs. Average	Range*	Body Build†	Estimated Average Weight, lbs. of Newborn Foal	Remarks
1. American Saddle Horse	62.0	60–65	1000	850–1150	0.842	99	3- and 5-gaited show horse
2. Roadster (light carriage)	62.0	60–66	1010	850–1150	0.850	100	Standardbred show horse
3. Colorado Ranger	64.5	63–66	1150	1000–1300	0.859	111	Leopard-spotted coat pattern
4. Arabian	59.5	56–63	910	800–1050	0.862	91	Ancestral Oriental horse
5. Pinto	61.5	58–65	1000	800–1200	0.862	99	cf. American Paint
6. Standardbred	63.0	60–66	1060	950–1200	0.865	104	Trotters and Pacers with sulky
7. Mustang (feral)	55.0	52–58	720	600– 850	0.869	75	Western Plains pony
8. Tennessee Walking Horse	63.0	60–66	1100	950–1250	0.883	107	3-Gaited show horse
9. American Paint	60.0	56–64	950	780–1150	0.883	94	cf. Pinto (both are piebald or skewbald)
10. Palomino	62.0	60–64	1050	900–1200	0.884	103	Golden parade horse
11. Thoroughbred	65.0	62–68	1200	1050–1350	0.901	115	Aristocrat of racehorses
12. Hunter (heavyweight)	66.0	60–72	1300	1000–1600	0.903	124	To carry rider of up to 205 lbs. (Middleweight Hunter comes between)
13. Hunter (lightweight)	64.5	64–66	1200	1100–1300	0.905	115	To carry rider of up to 165 lbs.
14. Polo Pony	61.0	58–64	1050	900–1200	0.928	103	Usually about ⅞ Thoroughbred
15. Anglo-Arab (Half-Bred)	60.0	57–63	1000	850–1150	0.929	99	Arab x Thoroughbred
16. Pony of the Americas	50.0	46–54	580	450– 750	0.931	63	Appaloosa coloration
17. Shetland, American type	44.0	42–46	400	350– 460	0.943	47	More slender than Purebred; often piebald or skewbald
18. American Buckskin	60.5	58–63	1050	900–1200	0.951	103	Largely from Spanish and Norwegian duns.
19. Morgan	59.7	56–64	1010	800–1200	0.952	100	Closest breed to *medium* size and build
20. Appaloosa	61.6	58–65	1120	950–1300	0.960	108	The third most popular U. S. breed
21. Welsh Pony	50.0	47–53	600	500– 715	0.963	64	Several types are recognized
22. Appaloosa Pony (over 48 in.)	52.0	48–56	675	540– 840	0.965	71	A smaller edition of the Appaloosa horse, with similar markings
23. Appaloosa Pony (under 48 in.)	44.0	44–48	410	315– 530	0.968	48	
24. Andalusian	61.0	59–63	1100	1000–1200	0.972	107	Spanish-Barb cross
25. Hackney Horse	61.0	59–63	1100	1000–1200	0.972	107	A natural trotter, now much used in show and jumping
26. Gotland Pony	49.0	46–52	570	475– 680	0.975	62	Known also as Viking Pony
27. Hackney Pony (over 52 in.)	53.0	52–54	725	685– 770	0.975	75	High-stepping show pony
28. Hackney Pony (under 52 in.)	50.7	50–52	635	600– 675	0.976	68	High-stepping show pony
29. Galiceno	51.0	48–54	660	550– 770	0.983	69	From Galicia, Spain
30. Cleveland Bay	65.0	63–67	1440	1250–1650	1.051	136	Often crossed with Thoroughbred to produce Hunters
31. Quarter Horse	59.0	57–61	1150	1040–1300	1.121	106	Most popular breed in U.S.
32. Percheron	64.5	60–70	1540	1200–2000	1.151	145	Most numerous draft breed in U.S.
33. Shetland Pony, purebred	39.6	38–42	365	310– 420	1.179	44	The best specimens are of solid coloration; black, brown, or bay
34. Belgian	64.5	60–70	1600	1200–2100	1.196	150	The fastest-growing in numbers of draft breeds in the U.S.
35. Clydesdale	65.0	60–72	1650	1250–2150	1.205	154	Of Scotch (county Lanark) origin
36. Shire	66.0	60–72	1750	1350–2300	1.221	163	Generally considered the largest draft breed
37. Suffolk	64.0	60–69	1600	1200–2000	1.224	150	Chestnut only in coloration

* The ranges given here are those ordinarily prevailing among 50 to 100 horses. Maximum ranges would in most breeds be approximately *double* the plus or minus deviations shown in this table. A Thoroughbred standing 73 inches high, as did Limbo, would be expected to occur only once in about 2,000,000 racehorses.
† The formula for Body Build used herein is: Build = 206.6 x Body Weight, lbs./Height, in.³ This is for male and female horses *combined*. For males the multiplier is 201.3; for females, 199.9.

Referring again to Table 18, some so-called breeds, such as the Albino, which is rather a color type and includes horses of all sizes from ponies to drafters, obviously cannot be listed with respect to a *typical* size. Again, in certain breeds, such as the Hackney and the Welsh pony, two or more distinct types, so far as size is concerned, are included under the breed name. And even among types, as for example the Hunter, there are differing sizes for various weights of riders. On the other hand, in certain breeds, especially those used primarily in the show ring, often a *limited* range in size is desired in order to preserve distinct characteristics. Some breeds exhibiting this size-restriction are Hackney horses and ponies, purebred Shetlands, Andalusians, and Cleveland Bays.

In our table the Pinto and the American Paint Horse are listed separately, although the two are commonly regarded as being one and the same color type. If so, the slight size-difference here listed may be disregarded and the size-range extended to include the horses of both names.

While Shetland stallions and Percheron stallions are of about the same build (1.179), the *average* build when mares are included is higher in the Shetland on account of there being very little difference in build between the sexes, whereas in draft horses the difference may be considerable.

Actually, so far as physique is concerned, there are only three well-recognized types of the domestic horse: light breeds, draft breeds, and ponies; and even among these types there are overlappings in which no distinct separation as to size or conformation can be demonstrated.

In Table 18 the weights of the newborn foals are estimated from the formula given previously on page 44, namely: bodyweight of newborn = .0854 adult weight (of same sex) + 13.15 pounds. In the table the bodyweights of the sire and the dam taken together are *averaged* for this purpose. It will be noted that, as the formula indicates, the *smaller* the adult horse the relatively *larger* the newborn foal. This rule appears to prevail in various orders of mammals: that the size of the newborn animal varies less than that of the adult parents. A wide range exists in the birth weights among all breeds of horses and ponies; however, the weights given in Table 18 may be accepted as optimum, being neither below par nor excessive.

If all 37 breeds or types of the horses and ponies listed in Table 18 are averaged, the height is 58.50 inches and the weight 966.5 pounds. If the 11 breeds of ponies (whose average height is

49 inches and weight 566 pounds) are exluded, the averages for a full-sized horse (including both light and draft types) become: height 62.53 inches, weight 1180.5 pounds. Most saddle horses range between 59 and 65 inches, averaging 62.0, and 900 to 1200 pounds, averaging 1040. These figures correspond closely with those prescribed for U.S. Army cavalry horses, which are: height 61.5 inches (60–63), and weight 1020 pounds (950–1100). In our series, the horse that corresponds most closely with the mean in both *size* and *body build* is the Morgan.

9
Optimum Weekly Gains in Chest Girth and Body Weight from Birth to Five Years

WHILE GROWTH IN THE LENGTHS OF ALL THE LONG (LIMB) BONES, and consequently such external measurements as withers height, leg length, and trunk length, normally cease at about five years of age, the *thickness* of the bones, as indicated in the girths of the cannons, the shoulder width, and the croup (hip) width, may continue throughout the lifetime of the animal. For all practical purposes, however, a horse may be considered as mature or "fullgrown" by the age of five years. Approximately one-half (49.3 percent) of the body weight at five years is gained, on the average, by the age of one year, 26.6 percent between one year and two years, and the remaining 24.1 percent between two years and five years. This, however, does not mean that further increases in girth and weight may not occur *after* the age of five years. One has only to check the listed measurements of draft horses entered in the usual horse shows and livestock expositions to note that the sizes of "aged" stallions and mares are greater in almost every respect than those of three- and four-year-olds.

The theoretical figures given here in Table 19 have been derived from the percentage growth increases shown in Tables 5 and 6, respectively. These (Table 19) figures are simply a conversion of the latter monthly or multimonthly increases in *percentages* to absolute gains in *inches and pounds per week*.

The calculations in Table 19 to two decimal places for weight and three decimal places for chest girth were carried to this degree of precision simply in order for the total gains from birth to sixty

TABLE 19

Optimum *Weekly* Gains in Chest Girth (top rows) and Body Weight (bottom rows)
Figures apply to all breeds and both sexes.

Age Interval	Chest Girth, in., at Birth (top row) and *Body Weight, lbs., at Birth (bottom row)*										Relative Increase	
	22.00 *40*	23.38 *47*	26.20 *64*	28.55 *81*	30.72 *99*	32.54 *116*	34.18 *133*	35.69 *150*	37.09 *167*	38.39 *184*	Chest Girth	*Body Weight*
Birth – 1 mo.	0.847 *4.11*	0.973 *5.58*	1.146 *8.48*	1.283 *11.37*	1.386 *14.26*	1.476 *17.15*	1.556 *20.04*	1.627 *22.94*	1.693 *25.83*	1.754 *28.72*	1.0000	*1.0000*
1 mo. – 2 mo.	0.567 *3.62*	0.651 *4.91*	0.767 *7.46*	0.859 *10.01*	0.928 *12.56*	0.989 *15.11*	1.042 *17.65*	1.090 *20.20*	1.134 *22.74*	1.175 *25.29*	0.0086	*0.8805*
2 mo. – 3 mo.	0.427 *3.13*	0.491 *4.25*	0.577 *6.45*	0.647 *8.65*	0.699 *10.85*	0.745 *13.05*	0.785 *15.25*	0.822 *17.45*	0.855 *19.65*	0.886 *21.85*	0.5046	*0.7609*
3 mo. – 4 mo.	0.313 *2.65*	0.360 *3.60*	0.424 *5.46*	0.475 *7.33*	0.513 *9.19*	0.546 *11.06*	0.575 *12.93*	0.603 *14.79*	0.628 *16.66*	0.651 *18.53*	0.3693	*0.6446*
4 mo. – 5 mo.	0.255 *2.35*	0.294 *3.19*	0.345 *4.84*	0.387 *6.50*	0.418 *8.16*	0.445 *9.82*	0.468 *11.47*	0.491 *13.12*	0.511 *14.77*	0.529 *16.42*	0.3016	*0.5719*
5 mo. – 6 mo.	0.216 *2.17*	0.248 *2.94*	0.292 *4.46*	0.328 *5.98*	0.354 *7.50*	0.377 *9.03*	0.397 *10.55*	0.417 *12.08*	0.434 *13.60*	0.450 *15.12*	0.2557	*0.5267*
6 mo. – 7 mo.	0.197 *2.06*	0.226 *2.79*	0.267 *4.23*	0.299 *5.67*	0.323 *7.12*	0.344 *8.56*	0.362 *10.01*	0.380 *11.45*	0.396 *12.90*	0.412 *14.35*	0.2328	*0.4992*
7 mo. – 8 mo.	0.188 *1.99*	0.215 *2.70*	0.255 *4.10*	0.286 *5.50*	0.309 *6.89*	0.329 *8.29*	0.347 *9.68*	0.364 *11.08*	0.379 *12.47*	0.393 *13.87*	0.2220	*0.4830*
8 mo. – 9 mo.	0.180 *1.94*	0.206 *2.63*	0.244 *4.00*	0.273 *5.36*	0.295 *6.73*	0.314 *8.09*	0.331 *9.46*	0.348 *10.82*	0.362 *12.19*	0.377 *13.55*	0.2131	*0.4717*
9 mo. – 10 mo.	0.174 *1.90*	0.199 *2.57*	0.236 *3.91*	0.264 *5.25*	0.285 *6.59*	0.303 *7.92*	0.319 *9.26*	0.335 *10.60*	0.350 *11.94*	0.364 *13.27*	0.2053	*0.4620*
10 mo. – 11 mo.	0.169 *1.86*	0.193 *2.52*	0.229 *3.83*	0.256 *5.14*	0.276 *6.45*	0.294 *7.76*	0.310 *9.07*	0.325 *10.38*	0.339 *11.69*	0.352 *13.00*	0.1990	*0.4523*
11 mo. – 12 mo.	0.164 *1.82*	0.188 *2.47*	0.222 *3.75*	0.249 *5.03*	0.269 *6.31*	0.287 *7.59*	0.302 *8.87*	0.316 *10.15*	0.329 *11.43*	0.341 *12.71*	0.1938	*0.4425*
12 mo. – 15 mo.	0.155 *1.68*	0.178 *2.28*	0.210 *3.46*	0.235 *4.64*	0.254 *5.82*	0.271 *7.00*	0.286 *8.18*	0.299 *9.36*	0.311 *10.54*	0.322 *11.72*	0.1827	*0.4082*
15 mo. – 18 mo.	0.125 *1.37*	0.145 *1.87*	0.169 *2.84*	0.189 *3.81*	0.204 *4.78*	0.217 *575*	0.229 *6.72*	0.239 *7.68*	0.249 *8.65*	0.258 *9.62*	0.1476	*0.3349*
18 mo. – 21 mo.	0.076 *1.17*	0.087 *1.58*	0.103 *2.40*	0.115 *3.22*	0.124 *4.04*	0.132 *4.86*	0.139 *5.68*	0.145 *6.50*	0.151 *7.32*	0.156 *8.14*	0.0895	*0.2833*
21 mo. – 24 mo.	0.037 *1.11*	0.043 *1.50*	0.050 *2.27*	0.056 *3.05*	0.060 *3.82*	0.064 *4.60*	0.067 *5.37*	0.070 *6.15*	0.073 *6.93*	0.076 *7.71*	0.0443	*0.2682*
24 mo. – 30 mo.	0.030 *0.93*	0.034 *1.26*	0.041 *1.92*	0.045 *2.57*	0.049 *3.23*	0.052 *3.88*	0.055 *4.53*	0.057 *5.18*	0.059 *5.83*	0.061 *6.48*	0.0350	*0.2256*
30 mo. – 36 mo.	0.024 *0.61*	0.027 *0.84*	0.033 *1.27*	0.036 *1.71*	0.039 *2.14*	0.042 *2.57*	0.045 *3.00*	0.047 *3.43*	0.049 *3.86*	0.051 *4.29*	0.0287	*0.1497*
36 mo. – 42 mo.	0.019 *0.42*	0.021 *0.57*	0.026 *0.87*	0.029 *1.16*	0.031 *1.46*	0.033 *1.75*	0.035 *2.05*	0.037 *2.34*	0.039 *2.63*	0.040 *2.92*	0.0220	*0.1020*
42 mo. – 48 mo.	0.014 *0.24*	0.016 *0.32*	0.019 *0.49*	0.021 *0.66*	0.023 *0.83*	0.024 *1.00*	0.025 *1.17*	0.026 *1.34*	0.027 *1.51*	0.028 *1.68*	0.0172	*0.0582*
48 mo. – 54 mo.	0.007 *0.12*	0.008 *0.16*	0.009 *0.25*	0.010 *0.33*	0.011 *0.42*	0.012 *0.50*	0.013 *0.59*	0.014 *0.67*	0.014 *0. 76*	0.015 *0.84*	0.0086	*0.0293*
54 mo. – 60 mo.	0.003 *0.09*	0.004 *0.12*	0.004 *0.18*	0.005 *0.24*	0.005 *0.30*	0.005 *0.36*	0.006 *0.42*	0.006 *0.47*	0.006 *0.53*	0.006 *0.59*	0.0032	*0.0210*
Adult Chest Girth, in.	45.66	50.53	58.23	64.36	69.41	73.75	77.64	81.18	84.43	87.45		
Adult Body Weight, lbs.	300	400	600	800	1000	1200	1400	1600	1800	2000		

NOTE: To derive the total amount to be gained between birth and 60 months (in either chest girth or body weight), multiply each weekly increase from birth to 12 months inclusive by 4 1/3; from 12 months to 24 months inclusive by 13; and from 24 months to 60 months inclusive by 26; then add the three products. For example, with a mature body weight of 600 pounds, the birth weight is approximately 64 pounds, which leaves a remainder of 600—64, or 536 pounds, to be gained from birth to 60 months. The 536 pounds is distributed according to the *weekly* increments listed in the bottom rows in the table under 64 pounds birth weight.

months to add up consistently. For all practical purposes the deci-
mals can be ignored. For example, in a horse of near-average size,
which would be expected to weigh at five years 1000 pounds and
which would start accordingly with a birth weight of about
100 pounds, the optimum gain in weight between one and two
months of age would be *12 or 13 pounds a week,* and from six to
eight months *about 7 pounds a week,* and so on. The chest girth during the same periods should increase, as shown in the table, at the
rate of about .9 of an inch a week between one and two months and
about .3 of an inch a week between six and eight months.

The seventeen-pound increases in the birth weights result from
the birth weight normally varying by 8.54 pounds for every 100-
pound variation in adult weight (and so, about 17 pounds for each
200-pound adult body weight increase shown in the table). The
chest girths, both at birth and in the adult, show no such uniform
variation, on account of being related to the *cube root* of the body
weight, which yields an irregular increase.

Our table shows the decidedly *different rates* of growth applying to chest girth and body weight, respectively. During the first
year, chest girth gains 67.75 percent of its total gain from birth to
five years; from one year to two years, 21.60 percent; and from two
years to five years, 10.65 percent. In contrast, these percentages for
body weight are 49.3, 26.6, and 24.1, respectively. These differing
growth rates are shown graphically in Figure 2.

"Puberty" in a horse may be regarded as the stage at which the
body weight is approximately two-thirds (68 percent) that at maturity. The relative age of a horse does not, as is commonly stated,
vary as a uniform fraction of the age of man. Rather, the horse's
age varies approximately as $\frac{1}{2}$ the age of man, minus $5\frac{1}{2}$ years. A
horse therefore reaches puberty at about $1\frac{1}{2}$ years, maturity at
5 years, and extreme old age at 53 or 54 years, these respective ages
corresponding with 14 years, 21 years, and 117–119 years in man.
On this scale, taking the *average* age at death of man to be 75 years,
that of a horse should be 32 years. As the latter figure is about
5 years greater than that usually assumed, it may be that most horses die prematurely rather than from true "old age."

The age at which the birth weight of the foal is doubled varies
inversely according to the birth weight. According to our table, a
40-pound pony foal should take about $10\frac{2}{3}$ weeks, or 75 days, to
double its birth weight; a 99-pound foal, about $7\frac{1}{4}$ weeks, or 51 days;

and a 184-pound draft foal, about 6⅔ weeks, or 47 days. However, there is a wide range of variation in this respect, according to the degree of nutrition, and some foals double their birth weight by the end of the first month. But this may be the result of prior malnutrition followed by overfeeding, rather than evidence of *normal* growth.

10
Review and Summation
Of Part I

BODY MEASUREMENTS OF NUMEROUS BREEDS AND TYPES OF HORSES ARE given in 73 previous studies which are herein reviewed.

In Tables 1 and 2, measurements of height, chest girth, cannon girth, and body weight are given for 17 male light breeds, 16 female light breeds, 5 male draft breeds, 3 female draft breeds, and for male and female Shetland ponies. In these 44 series abstracted from various authors, 3265 male and 1640 female horses are represented. From both tables, formulas for the estimation of body weight from chest girth have been derived.

In Tables 3, 4, 5, and 6 are given the typical or optimum rates of growth from birth to five years in withers height, cannon girth, chest girth, and body weight, as derived from six to eight series of different breeds.

In Table 7, body weight and cannon girth are related to chest girth, the latter ranging from 20 inches to 55 inches in Shetland and other ponies, 30 to 78 inches in light horses, and 30 inches to 100 inches in draft horses. From this table can be derived body weight from chest girth in both sexes of any breed, from birth to adult.

In Table 8 the information given in Table 7 is expanded so as to show the proportionate or expected body weight at any age from one month to five years, as based on the weight at birth. This table accommodates all sizes of ponies and horses from Shetlands to 2000-pound drafters. The same range in body weight is shown also on a graph (Fig. 4), on which may be plotted and checked the growing foal's weight at various ages.

In the text of these pages and in Tables 9 to 16 inclusive are given detailed body measurements of adult male and female horses of the following breeds: Arab, Thoroughbred, Standardbred, Quarter Horse, Morgan, Appaloosa, draft, and Shetland, respectively. In addition, for the Arab, Thoroughbred, Quarter Horse, draft, and Shetland, measurements of newborn male and female foals are included. Some information is given also on the Tennessee Walking Horse and the American Saddle Horse.

In Table 17 are shown the characteristic sex differences in body measurements and proportions in adult light horses. These differences are discussed on pages 87–89 of the text.

A comparison of the physique of the newborn foal is made with that of the adult horse, using the measurements listed in Tables 9 (Arab), 10 (Thoroughbred), 12 (Quarter Horse), 15 (draft), and 16 (Shetland).

Heights, weights, and other data are given on Man o' War and other famous racehorses, along with information on the pulling powers of draft horses and ponies. The latter are discussed also as to their relative speed in trotting races.

A discussion is made of Table 18, in which are listed the typical heights and weights of 37 breeds and/or types of horses and ponies popular in the United States today.

In Table 19 are estimated the optimum weekly gains in both chest girth and body weight, from birth to the adult age of five years.

On the basis of the information given in Part I on the weights and other body measurements of horses of various breeds from birth to adult, it is now opportune to describe and prescribe how the feeding of the growing foal should be regulated so as to promote optimum nutrition.

Part II
Nutrition

"Scientific" nutrition, either in man, farm animals, or zoo animals, can be an exacting and many-faceted procedure. My purpose here is to avoid technicalities, which may be of interest only to specialists, and to present, as simply as possible, a synthesis of the findings of nutritional experts whose writings are directed to the attention of ordinary, everyday horsemen and practical breeders.*

---

* Some of the leading books, papers, and breeders' bulletins consulted in this connection are listed in the Bibliography on nutrition.

11
The Relation of Nutrition to Growth

THIS RELATIONSHIP APPLIES, OF COURSE, TO ANIMALS OF ALL KINDS. *Nutrition* (from the Latin *nutrire,* to nourish) has been defined as the sum of the processes by which an animal absorbs, or takes in and utilizes, food substances. The term covers such processes as acquisition, alteration (as by mastication), digestion, assimilation, metabolism, and excretion of the nutriment essential to existence, growth, development, and reproduction. These processes, which together are known as *bioenergetics,* are in turn influenced by such factors as hunger (appetite) and repletion; activity of the central nervous system; basal metabolism; specific dynamic action; physical (muscular) activity; the use of proteins, fats, carbohydrates, minerals, vitamins, and water in bodily growth, maintenance, and reproduction; and the storage of surplus energy in the form of adipose tissue.

All energy derived and used by an animal ultimately comes from the rays of the sun. Much of this acquired energy is lost through respiration (exhalation) and in such ejecta as feces and urine. Most of the energy derived from food is used to maintain the protoplasm of the body's cells, which have the power not only of appropriating materials from the blood, but of utilizing these materials to maintain their own particular structures and functions.* The assimilative or upbuilding process is termed *anabolism;* the disassimilative or breaking-down process, *katabolism;* and the sum of the two processes, *metabolism.* The factors affecting metabolism and food intake include: age, sex, the automatic processes of circulation, respiration, excretion, and tonus, as well as such voluntary muscular activities as are involved in standing, walking, running, working, etc.

---

* For readers who desire to research some of the earlier literature on energy metabolism, a reference list of over 800 titles is presented by Samuel Brody in Research Bulletin 143 (June, 1920) of the College of Agriculture of the University of Missouri.

All the phenomena of growth and development are the results of the assimilative (anabolic) processes by which living animals are able to build their bodies from dissimilar substances, as when an animal subsists or grows by consuming vegetables or other inanimate materials.

Development is the process by which each tissue or organ of an animal's body is either formed (built up) or, if already incompletely formed, is so modified in shape and structure as to be fitted for a function of a higher order. The changing of a subadult animal, which may already be as large or nearly as large as its parents, to the fully-completed or mature stage is an example of development.

Growth, on the other hand, is the simple increase in size of a part, or of the entire body, by the addition of materials similar to those of which it is already composed. Growth and development, accordingly, may proceed together, with either process continuing after the other has ceased.

Nutrition, to repeat, is the process by which the various systems of the body are maintained in their normal form, size, composition, and state of functioning. In order that nutrition—particularly, optimum nutrition—may proceed, certain conditions must be present. First, the blood must be normal in composition and amount and circulate with suitable rapidity. Then there must exist a certain level of nervous stimulation and control; and the part to be nourished must have the power to appropriate the materials brought to it by the blood. In the disorder known as anemia ("lacking blood"), the carrying capacity of the blood is lowered and nutrition correspondingly impaired. And when the normal nerve supply to a part is reduced or cut off, atrophy or even complete loss of function in that part may result. Such a condition may occur if the spinal cord is injured or becomes diseased.

It is not only necessary that an animal's body be supplied with food in order that its natural functions may be performed, but it is equally essential that the food supplied should consist of the proper materials. In the horse, the requisite proportions of the food elements needed to maintain optimum nutrition are given in the pages that follow. Before dealing with this subject, however, it will be appropriate to point out the distinctive nature of the horse's digestive system.

12
The Digestive System
of the Horse

ONE OF THE FIRST ESSENTIALS IN THE FEEDING OF A HORSE IS TO UNderstand how this animal's digestive tract—and so the manner in which it must be fed—differs from that of other farm animals, such as the cow (with which, in general size, the horse may be compared). Figure 19 illustrates some of the differences between the digestive tracts of these two herbivorous, ungulate mammals.

The alimentary canal in an adult horse of average size consists of a tube over 100 feet in length that extends from the mouth to the anus and is modified along its length into an esophagus, small intestine, caecum, large intestine, and rectum, respectively (see Fig. 19). The great length of the digestive tract in the horse results mainly from the length of the small intestine, which is made more compact by means of its numerous folds or twistings. In the small intestine a considerable amount of digestion takes place, the material being acted upon principally by bile and pancreatic juice.

In Table 20 are compared the digestive capacities of the horse and the cow, each animal being presumably of average size.

The digestive tract of the horse is different both in structure and function from that of the cow, which is a ruminant, or cud-chewer. The horse's tract is much smaller; as a consequence, it cannot eat as much roughage (e.g., hay, straw, silage) as a cow. In the horse the beginning of the alimentary canal, or the esophagus, is relatively small in diameter and is surrounded by a strong, horseshoe-shaped band of muscular fibers. This construction causes the horse to have great difficulty in vomiting, or even in belching. Consequently, whatever the horse eats he must keep going in the one direction.

In the cow and other ruminants, the rumen (or first of its four stomachs) is positioned in front of, or anterior to, the small intestine, while in the horse the small intestine is followed by the

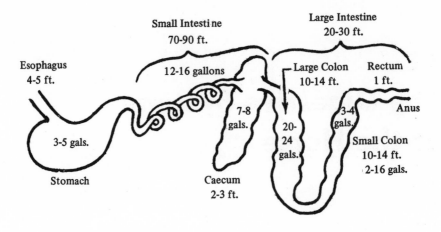

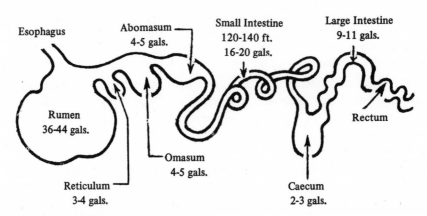

Fig. 19. *Schematic representations of the digestive tracts of the horse (above) and the cow (below), to show structural and capacity differences. See also Table 20.*

caecum, or "blind gut," which in turn precedes the large intestine. The caecum is of conical form, two to three feet long, and nearly a foot in diameter. Its base is near the lower part of the abdomen, while its apex is directed forwardly toward the chest. As a result of these structural and volumetric differences, the horse should be fed less dry, coarse roughage, more and higher-quality protein (no urea), and added B vitamins.

As to digestion, as mentioned above, much takes place in

TABLE 20

Parts and Capacities of the Digestive Tracts of Horse and Cow

Part of digestive tract	Capacity, gallons		Length, feet	
	horse	cow	horse	cow
Stomach	3–5	60–70	2	
Small intestine	12–16	16–20	70–90	130
Caecum	7–8	2–3	2–3	
Large intestine	30–35	9–11	20–30	
Total capacity, gallons	52–64	87–104	——	——

the small intestine, the walls of which are also a major source of absorption. Additional digestion, especially of cellulose, occurs in the caecum, but is relatively slight in amount. By the time the foodstuffs have passed through the colon they are fairly digested, and are then formed into balls of dung, or manure. Under natural conditions in the wild, horses and other Equidae graze almost continuously, taking in small amounts of food over a long period. This should be a key to their feeding as controlled by man.

13
Description of Nutrients

BY THE TERM NUTRIENT IS MEANT ALL FOOD ELEMENTS THAT PROVIDE nourishment to an animal. These elements may be broadly classed as protein, fat, carbohydrates, minerals, vitamins, air, and water. Their proper use provides the materials and energy needed to maintain the body, and where necessary to enable growth and development, reproduction, lactation, physical activity, and (as in the case of animals used for food) fattening to take place.

By the term *digestible nutrient* is meant that portion of a nutrient which may be digested and assimilated. What is known as a *balanced ration* is an amount of essential nutrients sufficient to nourish a given animal for a period of twenty-four hours.* *Total digestible nutrients* (TDN) is a term meaning the daily requirement of all foodstuffs (roughages, concentrates, and supplements), but not including water.

---

* While some writers on the subject question the need for "balanced" rations on the grounds that such balancing may be uneconomical if high-cost rations are used, all agree that if the feed ordinarily given is lacking in some essential ingredient, that ingredient (especially in the case of minerals and vitamins) must be supplied additionally if the animal's health and productivity are not to be impaired. (A full elucidation of this important subject is given in Chapter 11 of Morrison's *Feeds and Feeding*, 22nd edition, 1959.)

14
Nutritional Elements and Requirements

THE DAILY NEEDS FOR FOOD IN THE HORSE RANGE RATHER WIDELY, in accordance with age, sex (as in the case of a pregnant or lactating mare), size, and the amount of work performed by the individual animal. In striving for optimum nutrition, it is desirable that each horse's energy and feed requirements be evaluated *separately*. It is now opportune to consider in order each food element as mentioned above.

Protein

All animals, herbivorous as well as carnivorous or omnivorous, need proteins in a sufficient amount to build up and maintain the various structures of the body, such as the muscles, bones, ligaments, tendons, hair, skin, hoofs, blood, and internal organs. In horses, the protein content of the body ranges from about 10 percent in fat individuals to 20 percent in thin ones. As protein is composed chiefly of amino acids, the value of a particular form of protein is judged primarily by its amino acid content, which varies considerably among the numerous plant and animal sources of protein. At present, some twenty-three differing amino acids are recognized. They are commonly regarded as the "building stones" of proteins. According to Ensminger (1969, p. 363), about half the number of amino acids presently known are dispensable, and the other half indispensable.

In the horse, the *utilization* of amino acids is less efficient than in ruminating farm animals such as the cow, sheep, and goat. In addition, the caecum of the horse is situated beyond or posteriorly to the small intestine, the latter organ being the main area for the digestion and absorption of nutriment. For these reasons, it is recommended that the protein ration of horses be adequate in amount and contain high-quality amino acids. In this connection it is advantageous, just as it is in humans, to provide a *widely varied* diet. While the amino acids in the various available sources of protein may not be of the same quality, it has been found that deficient proteins, if combined with superior ones, will often result in a mixture that provides better nutrition than either protein used singly. However, since the need in horses for protein is relatively low, it can readily be met with the daily rations prescribed later.

Fat (lipids)

Fats and oils, like carbohydrates (sugars and starches), are composed mainly of carbon, hydrogen, and oxygen. These two classes of foodstuffs provide the major portion of a horse's energy ration. The daily amount of fats and carbohydrates that a horse requires depends upon the size of the animal and the amount of work he performs. If a horse is fed more of these food elements than he needs, because of his being idle, the excess is quickly transformed into adipose tissue, just as in a human being, and the animal becomes fat, overweight, and physically sluggish. Since fats contain a higher proportion of carbon and hydrogen than do carbohydrates, the fats when digested liberate more heat, furnishing some $2\frac{1}{4}$ times as much heat or energy per pound as do carbohydrates. Estimations of the fat content of a horse's body indicate that it increases with age, being as little as 2 percent in a newborn foal to 20 percent in an adult horse in good condition. The latter ratio may be compared with the 19 percent of fat in the average young man, although in the average young woman the content is much higher, about 28 percent, owing largely to the normally thicker layer of subcutaneous fat in women. But variation between the sexes in horses is far less than in humans.

Although a small amount of fat is desirable in the rations of horses (since fat is the carrier of fat-soluble vitamins A, D, E, and K), most animal nutritionists feel that the ordinary mixed rations of well-tended horses contain a sufficient proportion of this element. That ordinary pasture grass, if consumed in sufficient quantities, can alone produce fatness is shown by the zebras of East Africa (*Equus böhmi*), which at all times when grass is available present a plump, well-fed appearance.

Carbohydrates

The organic compounds embraced under this heading constitute about three-quarters of the dry matter in plants, this material being the main source of a horse's feed. As mentioned above, carbohydrates and fats are used as sources of heat and energy, and any excess of them is transformed into body fat. However, for supplying the normal energy needs of horses, carbohydrates are of prime importance. Ordinarily, these needs are met by feeding a liberal amount of grain and decreasing the amount of roughage. A deficiency of energy-producing feed, on the other hand, may hinder the growth of foals, and in mature horses result in underweight, poor condition, and lack of stamina.

Energy in the horse is required for maintenance, growth, reproduction, and physical activity or work. The amount of carbohydrates needed to produce an optimum level of energy depends on the size of the horse and the amount of work he is required to perform. A racehorse in action, for example, may use up to a hundred times the energy expended while at rest. If the amount of protein supplied in a horse's feed is more than is needed, the excess of it also produces energy, but at needlessly high expense. A given amount of a concentrate, such as grain, supplies about 1½ times as much digestible energy as a like amount of roughage. Since the capacity of a horse's digestive tract is limited, this is an important consideration. The fiber contained in growing pasture grass, whether fresh or dried, is a more digestible form of roughage than the fiber of most hay. Too, the fiber of hay cut early in the season is more digestible than that cut in the late-bloom or seed stages. Foals, draft horses, and racehorses, in particular, should have rations in which the general feeding rule of "low fiber and liberal grain" is observed.

Minerals

These important inorganic elements are essential for normal growth and development of the teeth and skeleton, and enter also into the composition of the blood and soft tissues. A factor that influences the amount of minerals in an animal's feed is the amount that was present in the soil where the grass and roughages were grown. For this reason the hay from certain areas may be so deficient in minerals that the latter have to be supplied additionally, as for example by blackstrap molasses or one of the alfalfa-molasses feeds. On the other hand, an excessive intake of certain minerals can be as harmful and serious as a deficiency. The chief minerals essential in a horse's feed are: calcium, phosphorus, sodium, chlorine, potassium, iron, copper, cobalt, iodine, manganese, zinc, sulfur, magnesium, selenium, and molybdenum. The latter ten substances, on account of the small proportions in which they are needed by the body, are called *trace minerals*. It may now be well to take a closer look at each of the foregoing fifteen minerals.

Calcium and Phosphorus

Calcium and phosphorus are required especially during early growth of the skeleton and dentition, and in lesser amounts for maintenance in adult horses. In animals where much grain but insufficient roughage is being fed, a shortage of calcium may exist; whereas if the animals are consuming mostly roughage but little grain (a more prevalent condition), the shortage is more likely to be in phosphorus than in calcium. In mares, during the last quarter of pregnancy, the need for both calcium and phosphorus is increased, as it is also during the period of nursing the foal. Too, aged horses may need anywhere from 30 percent to 50 percent more calcium and phosphorus than is required by younger horses. However, some authors state that excesses of these two elements in a horse's rations may lessen the utilization of other minerals, such as iron, magnesium, manganese, and zinc. For most horses the optimum ratio of calcium to phosphorus would appear to be about 1.1 to 1. Ensminger (1969, p. 377) states that 2:1 is acceptable if the calcium is derived from legumes. Where there is a shortage of calcium in a horse's rations, the needed amounts are withdrawn (resorbed) from the bones. And farm animals in general, including horses, are more likely to suffer from a lack of calcium and phosphorus than of any of the other minerals except salt.

Sodium and Chlorine

Sodium and chlorine, which together constitute common salt, are important in maintaining normal osmotic pressure within the body cells, upon which depends the absorption of nutrients and the excretion of waste products. Too, sodium assists in the production of bile, which aids in the digestion of fats and carbohydrates. Chlorine is important in the formation of the hydrochloric acid in gastric juice, and so in aiding the digestion of proteins. The amount of salt in an animal's system is markedly influenced by losses through perspiration; and since the horse is a profuse sweater, salt in addition to that contained in the forage should be supplied in pure form (granulated salt, block salt, or rock salt, depending on which form is cheapest or most convenient), especially to horses performing hard work and during hot weather. Two to three ounces is the recommended daily allowance of salt in a 1000-pound horse at ordinary work.

Magnesium

Magnesium is an essential element in the bodies of animals, although the amount needed is very small: about 1/9 ounce (3200 mg) per day in a 1000-pound horse. Most of the feeds given to horses contain an ample proportion of magnesium.

Potassium

This is another essential mineral element. In fact, an animal's body contains more potassium than it does of either sodium or chlorine. When forage plants, which are generally high in potassium content, constitute the main portion of a horse's feed, sodium may be excreted in a greater amount, in which case common salt may have to be given additionally. If potassium is lacking, it can be supplied in molasses or in oil meals as well (preferably) as by dry roughages (e.g., pasture grass, clover or alfalfa hay, oat straw, corn or sorghum stover, etc.), However, as in the case of most other minerals, the usual feeds generally supply adequate amounts of potassium. This is practically certain if the rations include 50 percent or over of forage.

Sulfur

Since this mineral is an essential component of most proteins, as well as of certain vitamins, it is necessary that it be included in the rations of animals, although the amount needed is very small. A sufficiency of sulfur is generally supplied by an ordinary, well-balanced feed ration. It would appear that if the protein requirement is met, the sulfur intake will be over 0.1 percent, in which case there is no advantage in adding sulfur to the ration.

Iodine

This element is essential in horses and other farm animals as well as in humans, since if there is an insufficiency of it the thyroid gland may enlarge to compensate for the lack, and so develop into a goiter. The horses most susceptible to an iodine deficiency are pregnant mares. Fortunately, the amount needed is, as in the case of sulfur, very small, and may be supplied in the form of iodized salt. In inland areas where the content of iodine in the soil may be low, additional iodine must be supplied to prevent the occurrence of goiter in humans and/or in farm animals. No benefit, however, will result from supplying more iodine than the body can utilize.

Cobalt

It has been found that an inadequate supply of cobalt can cause deficiency diseases in cattle, sheep, and swine, although horses are not thus afflicted. Except in areas where a cobalt deficiency is known to exist, or where it can be suspected from the symptoms shown, there is no advantage in supplying it additionally (as in salt or a mixture of minerals). If the animal shows a normal appetite for its feed, it is probable that it is getting sufficient cobalt. In horses, cobalt is required for the synthesis of vitamin B_{12} in the intestinal tract; and an insufficiency of cobalt or of B_{12} may result in anemia. In such a case the needed cobalt may be supplied by mixing $\frac{1}{2}$ ounce of dissolved cobalt carbonate into 100 pounds of common salt. Care should be taken that the mixing is thorough, so that the proportion of cobalt is fairly uniform throughout the salt.

Copper

In young growing animals especially, a small proportion of copper is essential for normal development of the bones and cartilages. The recommended daily allowance in a 1000-pound horse is 64 mg (about one grain). In certain areas the pasturage may be low in copper and therefore need supplementing. Two good sources are alfalfa hay and prairie hay.

Iron

The recommended daily allowance of iron in a 1000-pound horse has been estimated as 640 mg (about 10 grains). All milk, including that of mares, is deficient in iron and cannot be made up through feeding iron supplements such as ferric oxide and ferrous carbonate. While ferrous sulfate may prove effective, it is recommended that foals should be creep fed as soon as they are old enough. As with most other minerals, an excess of iron may be harmful.

Manganese

As so far no deficiency of this element has been found in horses, it is evident that the need for it is very small and is adequately supplied by ordinary feed rations.

Zinc

Although comparatively little is known about the physiologic effects of this mineral, it appears to be essential in small amounts for bone, cartilage, skin, and hair formation. And as hoofs are structurally akin to hair, skin, and nails, an adequate intake of zinc is essential for the growth and maintenance of these structures. Accordingly too, ample zinc will improve a horse's haircoat. The recommended daily allowance in a 1000-pound horse is 400 mg (about 6 grains).

Selenium

This mineral element evidently is not required in the feed of horses, and there are instances where grazing horses have been poisoned from forage containing excessive amounts of selenium. In areas where this has occurred (two places are Wyoming and South Dakota), the poisoning if chronic is called "alkali disease," and if acute, "blind staggers." Chronic selenium toxicity may result from rations containing as little as 5 ppm (parts per million), and acute poisoning may occur where the feed contains 500 ppm or more. These conditions are usually confined to localities where the soils contain from 0.5 to 40 ppm of selenium.

Molybdenum

No quantative requirement of this element in the horse has been demonstrated. Where a toxic condition has resulted from an excessive molybdenum intake (16 or more ppm), it may be counteracted by administering a quantity of copper that is double that of the molybdenum.

Vitamins

The function of these important organic compounds is to assist in the utilization of various nutrients. They are essential for normal metabolism, but cannot in most cases be synthesized by the body. However, many vitamins are formed by bacterial action in the intestines. Vitamins are separated into two major groups, known respectively as fat-soluble and water-soluble vitamins. The fat-soluble group comprises vitamins A, D, E, and K.* The water-soluble group is represented by the vitamin-B complex, which at present includes 20 or more identified vitamins. These are designated by name rather than given a letter in the order of their discovery, as is done in the fat-soluble group.

The effects of the presence or absence of vitamins in the body is either identical or similar in animals (specifically, mammals) of all kinds, including man. When a vitamin deficiency is present, it is nearly always of a multiple nature rather than of some particular vitamin. Herbivorous animals such as horses, cattle, and sheep must necessarily obtain their natural-source vitamins from vegetable matter; whereas man, and carnivorous animals such as dogs and cats, can derive certain vitamins from animal sources such as liver, kidney, cod-liver oil, etc. Of course, this does not mean that the nutrition of herbivorous farm animals need in any respect be deficient, since most vitamins are to be found in fresh green forages and good-quality hays, or in such supplemental extracts (even some of animal origin) as may be added to their foods. The following review is not intended to be exhaustive, but is confined to those vitamins the properties and effects of which are best known.**

Fat-soluble Group

VITAMIN A (AND CAROTENE) This most important vitamin is essential for the maintenance of all tissues exposed to the outside of the body, such as the eyes, skin, respiratory organs, digestive tract, and reproductive system—in fact, all mucous membranes in which infec-

---

* Vitamin C, the antiscorbutic vitamin, need not be considered here in connection with the horse.
** For highly detailed accounts of the vitamins, readers are referred to Morrison (1962) and Ensminger (1969). See Bibliography.

tions may occur. The vitamin A requirement in horses may be supplied either by carotene (the counterpart in plants of vitamin A, which is strictly of animal origin) or by the vitamin itself. Good sources of carotene are fresh green pasturage and/or high-quality hay. However, hay that has been stored for a year or more is generally lacking in vitamin A because of a decline (oxygenation) in the carotene content due to such prolonged storage. An insufficiency of vitamin A causes a hardening of certain tissues, as in the eye (the "night blindness" of horses), and in young animals a deficiency in growth. During periods of reduced forage, as in winter and early spring, horses are able to draw upon stores of vitamin A that are present in the body, mostly in the liver. Mature horses that have had ample pasturage earlier are thus able to store up as much as a six-months' supply of this vitamin. Foals, however, store much less. The recommended daily allowance of vitamin A in a 1000-pound horse is 50 milligrams or 50,000 U.S.P. units. Horses at hard work, on the race track, or in hot weather *may* have a higher vitamin A requirement than this, while in pregnant mares the need may be up to five times the minimum maintenance requirements.

VITAMIN D This vitamin acts to regulate the absorption and deposition of phosphorus and calcium in the skeleton. It is generally considered that the presence of vitamin D in horses is rarely deficient, since it is produced by the action of ultraviolet rays from the sun on oil in an animal's skin. This vitamin is also present in forages that have been cured in the sun. Too, vitamin D in stored hay does not deteriorate to the extent that does vitamin A. The recommended daily allowance of vitamin D in a 1000-pound horse is about 3000 U.S.P. units. A lack of vitamin D in foals may cause rickets; and in adult horses, osteomalacia (morbid softening of the bones) ; while an excess may result in extensive soft-tissue calcification. Toxicity from vitamin D is more likely to occur and be severe in proportion to the excess of calcium in the horse's feed. And because of the usual presence of vitamin D in mineral-vitamin supplements, horses are more apt to suffer from an excess of this vitamin than from a deficiency.

VITAMIN E (TOCOPHEROL) While there is still uncertainty as to all the properties of this vitamin, the consensus is that it protects the red blood cells and fat compounds of the body from destruction by oxygen, allowing the oxidizing agents to work on itself instead. In this way vitamin E enables vitamin A to work with greater effectiveness. But whether in the horse this results in greater fertility and improved performance has not been proved. Most rations provide adequate amounts of vitamin E. Hence, before adding any, it is best to check with a veterinarian.

VITAMIN K   This vitamin is involved in clotting of the blood. In the horse it would appear that it is synthesized in ample amounts by the bacteria in the intestines. However, if the blood is unduly slow in clotting, a deficiency in vitamin K may be suspected.

Water-soluble Group

THIAMINE (VITAMIN B₁)   The requirement by horses for this vitamin has not been established. However, three parts in a thousand (= 3000 ppm) of the feed intake is said to be adequate for maintenance, body weight gains, and normal levels of thiamine in the skeletal muscles. The vitamin is synthesized in the caecum and intestines by bacterial action, and about 25 percent of the amount thus produced may be absorbed. Horses fed on hay and grass of poor quality may develop a thiamine deficiency, the symptoms of which are a loss of weight, nervousness, poor coordination of their limb movements, and hypertrophy of the heart. Such animals should have thiamine added to their feed, as should also all horses subjected to extra stresses, as in showing and racing. Horses that have been poisoned by consuming plants that contain antithiamines (as for example bracken fern and mare's-tail) may gain relief by being given periodic subcutaneous injections of 100 mg of thiamine.

RIBOFLAVIN (VITAMIN B₂)   This element, which was formerly called vitamin G, has been described as a growth-promoting factor. It appears to be essential likewise for proper nutrition at all ages. A deficiency of riboflavin is likely to cause inefficient digestion, poor structural development, general weakness, nervous symptoms, lowered resistance to disease, and the eye disorder known as "moon blindness" (periodic ophthalmia). However, the last-named affliction often follows leptospiral infection. The recommended daily allowance of riboflavin in a 1000-pound horse is 20 mg (about 1/3 of a grain).

NIACIN   This vitamin, which is known also as nicotinic acid and pellegra-preventive, is essential for carbohydrate, fat, and protein metabolism. In horses it appears to be synthesized in adequate amounts by the bacterial flora of the caecum and intestines. Yet, since the horse can convert the amino acid tryptophan into niacin, it is essential that adequate niacin in the rations be supplied, since otherwise the horse will draw tryptophan out of his tissues to make up the deficit. The recommended daily allowance of niacin in a 1000-pound horse is 50 mg. (about ¾ of a grain).

FOLIC ACID (FOLACIN) This B-complex vitamin is so called because it is present in plant foliage. Accordingly, it is widely distributed in forages. It is believed that in horses this vitamin is probably synthesized in ample amounts in the intestines; but a deficiency of it may cause anemia. The recommended daily allowance in a 1000-pound horse is very small: 2.5 mg. or about $\frac{1}{25}$ of a grain.

PANTOTHENIC ACID As in many other vitamins of the B-complex, pantothenic acid appears to be adequately supplied in the horse by bacterial synthesis in the intestinal tract. As in riboflavin, a deficiency of it results in poor growth, lessened appetite, skin rashes, and nervous disorders. The recommended daily allowance of pantothenic acid in a 1000-pound horse is 60 mg (nearly one grain).

VITAMIN B$_{12}$ This vitamin is associated with animal protein sources, but is also to be found in grass. Its functions are to stimulate appetite, accelerate growth, and aid in feed utilization and normal reproduction. It possesses also an antianemia factor. The recommended daily allowance of B$_{12}$ in a 1000-pound horse is indeed a minute amount: 125 micrograms ($= \frac{1}{8000}$ gram).

Water

Although we shall here deal with water after the numerous other elements in nutrition have been discussed, actually water is the most vital of them all, since animals can survive longer without food than without water, a basic element necessary for the normal functioning of all the systems of the body. Excluding fat and the contents of the digestive tract, the composition of the animal body is approximately 75 percent water (plus 20 percent protein and about 5 percent mineral matter).

In the horse, which is a profuse sweater, it is important that an ample supply of clean water be made available 24 hours a day. The amount required is in close ratio to the amount of dry matter ingested. In addition to the quantity of water needed for normal maintenance of the body, the loss occurring through sweating and the extra amount needed in lactating mares may increase the requirement considerably. The daily intake of water in a mature horse averages from 4 to 10 gallons. It may be considerably more or considerably less than this, depending upon the work performed and the amount lost through sweating and in the urine and feces and the water vapor exhaled. Captain Hayes, an authority in such matters, lists an average of 8.5 gallons: morning 1.9, midday 3.4, evening 3.2.*

Working horses should be watered at least 3 or 4 times a day. If the horse is hot, he should be cooled first before being allowed to drink all he wants. During the cooling period he may be given occasional sips of water. Horses in pasture, or not being worked, should have access to good clean water. While some authors say that water may be given either before, during, or after feeding, this advice needs to be qualified. If, for example, a horse is watered immediately after being given grain, the grain may wash out of his stomach and possibly bring on an attack of colic. Horses feeding on moist forage such as green grass naturally require less water than those on dry feed.

Water is best supplied by automatic waterers. The temperature of the water should be at least 40 degrees F. and preferably 50 degrees. Hence in cold seasons the water may require heating. Too, heating of the watering tanks will keep the water free of ice. The waterers should be placed both in stalls and corrals. Horses that are obliged to drink water from stagnant pools, pot holes, or contaminated streams are subject to water-borne parasitic diseases. If a horse becomes over-thirsty from taking too much salt, he may drink an excessive amount of water. And as the horse's digestive capacity is limited, this may restrict his intake of solid nutrients.

---

* A horse when drinking takes from a cupful to nearly two cupsful of water on each swallow, and drinks at the rate of from 65 to 90 swallows a minute. Hence, if a horse is allowed to drink for a minute, the quantity of water imbibed would range from about 16 to 40 pints, or 2 to 5 gallons.

15
Some General Principles of Nutrition

UNDER NATURAL CONDITIONS, A HORSE CAN SUSTAIN HIMSELF BY FEEDING solely on the grass and other ground vegetation characteristic of his natural habitat, the open plains. This is true also of the wild forms of the horse family, some of which, such as the Mongolian wild horse and the kulan, live under harsh desert conditions of scanty food and water, along with extremes of heat and cold according to the season. Yet such equines are able to survive, reproduce, and even keep in fit physical condition. And where the vegetation at certain times of the year is profuse and lush, as occurs on the grasslands or savannahs of south and east Africa, the zebras even grow fat, despite all their activity in galloping away from carnivores.

In domestic horses likewise, good pasturage is the basis of natural nutrition. Kentucky and other varieties of bluegrass, all of which are rich in protein, have long been recognized as superior forage for horses. Alfalfa, whether green or as hay, is another good source, as are timothy and Bermuda grass, each of which has special qualities. To obtain the best nutrition from such grasses, however, they should be used when their quality is highest, which is generally after about four weeks of growth, whether this be the first growth in spring or the regrowth after hay harvesting. Also, if the pasture is known to be deficient in limestone and/or other minerals, adequate amounts of lime, phosphate, and possibly potash should periodically be added. Many authorities on animal nutrition give additional reasons why pastures should be used in preference to stall feeding, not the least of which is the lowered cost of feeding, which in some cases may be reduced by one half. But when pasturage is deficient either in amount or quality due to being snowed over or otherwise rendered unusable, alternative feeding in the forms of hay, grain, and mineral and vitamin supplements must be substituted. Some approved practices in connection with such supplemental feeding may now be considered. First, however, let us recapitulate the horse's manner of feeding.

I have previously (pp. 117–119) commented on the distinctive nature of the horse's digestive system and its limitations so far as capacity is concerned (see also Fig. 19 and Table 20). Owing to the horse's narrow esophagus (gullet), he has to masticate his feed thoroughly if he is to swallow it without difficulty.* If a horse bolts his food (as some horses have a habit of doing), the food may pass through his digestive tract without being digested and absorbed. In this respect horses are radically different from members of the ox tribe, which (especially if in the wild, and confronted with danger) can swallow their food practically unchewed and later rechew it at leisure. Basically, the horse's stomach is designed for small quantities of feed taken in over many hours' time, rather than large quantities at one time, as a ruminating animal such as the cow may ingest.

In the horse, the limited amount of digestion that takes place in the relatively small stomach is completed within 10 to 15 minutes. This may account for a horse—contrary to what he *should* do—being able to take in a large amount of food or water at one time. After the ingested material has left its brief stay in the stomach, it enters the small intestine, where much digestion takes place. Proteins are therein broken down into amino acids, starches into sugars, and so on for the absorption of certain other nutrients. In the caecum or "blind gut," where digestion continues, there is also some synthesis and absorption of vitamins, especially those of the B-complex. The capaciousness of the caecum, which is about four feet in length and a foot in diameter, enables the horse to take in rather large amounts of roughage, such as hay and forage. But in the caecum also, unfortunately, is where impaction and other digestive difficulties often occur. For all these reasons, horses, as compared with cattle, should be fed relatively less roughage and more high-quality proteins (but *not* nonprotein nitrogenous products such as urea), along with ample

---

* The chewing action of a horse is not unlike that of a cow or other ruminant, in which the lower jaw moves sidewise as well as up and down, in a sort of rotary motion. On account of the lower jaw being narrower than the upper, a horse can chew with his grinding (cheek) teeth only on one side of his mouth at a time. The number of chewing movements averages 75 per minute. For comparison, in man the number averages 90–100, in cattle 70, and in mice 350.

amounts of vitamins, especially vitamins K and those of the B-complex.

As to the quantity of feed a given horse requires, it is in proportion to the animal's age and size and the amount of daily work ordinarily performed. However, if we assume a given amount of activity (or inactivity), the amount of feed needed is in ratio to the body weight of the horse raised to the ¾ (0.75) power.* Significantly, the surface area of a horse's body is also closely in ratio to $BW^{0.75}$. Accordingly, the feed requirements are closely proportionate to the surface area of the horse. Another correlation, which I detected during the course of my researches on equine dentition, is that a horse's feed requirements are in ratio not only to the surface area of the external body, but also to the surface area of the *grinding teeth* (specifically, the area of one row of six upper cheek teeth). For example, the tooth area in a Shetland pony weighing 340 pounds was found to be 28.3 square centimeters;** in an Arab weighing 920 pounds, 41.7 sq. cm.; and in a draft horse weighing 1600 pounds, 51.5 sq. cm. The area of the teeth does not alter significantly with age, as might otherwise be expected; for as the *length* of the cheek-tooth row diminishes with age and wear, the *width* of the exposed tooth surface increases almost in proportion, so that a more or less uniform area for mastication is presented.

The tooth area (i.e., of one row of the upper cheek teeth) may be predicted from the formula: area, sq. cm. = 3.22 x body weight, lbs.$^{0.375}$.

The significance of all this is that the horse is a fitting example of the various *geometric* relationships that exist in the animal body, and which prevail not by chance but by the necessitated correspondences between length, breadth, and thickness, volume, surface, and weight. These relations in turn make it possible to prescribe the theoretically optimum amounts of feed required to nourish or sustain a horse of any size under any condition of physical activity. All that is necessary is to first determine the daily amount of feed required

---

* This is the generally-accepted correlation factor. However, in some mature working Percheron mares and geldings reported by Brody, Kibler, and Trowbridge (1943), the correlation is in *direct ratio* to body weight (i.e., $BW^{1.0}$). But this may be the result of an atypical growth rate in these draft horses, which grew too slowly between 6 months and 3 years and then too rapidly, without the characteristic slowing-down from about 3 years on.

** All measurements of the skeleton, skull, and teeth of an animal are taken and recorded in Metric units, i.e., either in centimeters or millimeters.

by a horse of a given size (body weight) in a state of idleness or rest, then modify that amount in accordance with the increased needs that result from increased activity or work.

Table 21, following, gives the proportionate values of body weight raised to the ¾ (0.75) and the 0.52 powers, respectively, which ratios express in the adult horse and the growing foal what is known as Metabolic Body Size and is the basis of all feed requirements listed in Tables 22 and 23.

TABLE 21

Relative Body Weight (RBW) raised to the ¾ (0.75) power (mature horses) and to the 0.52 power (growing foals), for the purpose of establishing the Metabolic Body Size, or basis for feed rations, etc.
(based on a 1000-pound Body Weight = 1.000)

Body Weight, lbs.	Relative $BW^{0.75}$	Relative $BW^{0.52}$
25	0.063	0.147
50	0.106	0.211
100	0.178	0.302
150	0.241	0.373
200	0.299	0.433
250	0.352	0.486
300	0.405	0.535
400	0.503	0.621
500	0.595	0.697
600	0.682	0.765
700	0.766	0.831
800	0.846	0.891
900	0.924	0.947
1000	**1.000**	**1.000**
1100	1.074	1.049
1200	1.147	1.097
1300	1.218	1.144
1400	1.288	1.190
1500	1.356	1.235
1600	1.423	1.279
1700	1.489	1.321
1800	1.554	1.361
1900	1.618	1.399
2000	1.682	1.454

16
Energy and Its Derivation

ENERGY HAS BEEN DEFINED AS THE CAPACITY FOR WORK. IT CAN BE IN reserve, that is, potential; or it can be applied in the performance of work and therefore become kinetic. It is also an expression of the production of heat by friction. Although early philosophers thought that energy, or force, could be destroyed, Leibnitz showed that in impact and in friction the energy manifested is converted into heat by the agitation of the particles of the body acted upon, thus being dissipated among the component parts of the body.

The doctrine of the Equivalence of Work and Heat is that either is transformable into the other; and the transformation takes place at a fixed rate. The amount of heat necessary to raise one kilogram of water one degree centigrade is called a "great" (or large) calorie. The latter is the calorie used in estimating human nutritional needs. The needs of farm animals are measured generally by the unit called a therm. However, the unit could just as well be the calorie, since a therm is equivalent to 1000 calories.

In animals, the production of work is due to those chemical transformations in the body which develop heat. The physiologist Jules-Robert Mayer, in a paper published in 1845, elucidated the process as follows:

> The chemical energy contained in the food consumed and the oxygen breathed is the source of two kinds of forces: movement and heat; and the sum of the physical forces produced by an animal is the equivalent of the total sum produced by the chemical process which has taken place at the same time. If all the mechanical work performed by a horse during a certain time after being converted into heat is added to the heat produced simultaneously in its body, the sum will be equal to the quantity of heat evolved in the corresponding chemical reaction.*

---

* This English translation of the original paragraph, which is in German, appears in *The Human Motor*, by Jules Amar (1920), p. 45.

The French physiologist Claude Bernard remarked: "The muscle is a machine in perpetual renovation"; and A. Chauveau added: "It is not what one is actually eating that furnishes the energy employed in the physiological work of the organism, but the potential energy formed by what has been eaten previously."*

In order for bodily action or labor to be performed, four conditions or elements must be present: heat (circulation or cumbustion), oxygen (air), moisture (water), and nourishment (food). The bodies of mammals in general contain about 60 percent of water, of which from 5 percent to 8 percent is lost daily through perspiration, urination, and other excretion. Thus a 1000-pound horse loses from 50 to 80 pounds of water daily, depending upon the amount of work performed and its effect on perspiration and other elimination.

As mentioned, the amount of nutrients needed to produce a given daily amount, or expenditure, of energy is customarily expressed in the units known as calories. One method of computing the number of calories in a given article of food (or feed) weighing 100 grams is to multiply the proportion (percentage) of dry matter in each of the chief constituents of the food by the following respective factors: carbohydrates 4.1, fates 9.1, proteids 4.1. For example, in 100 grams of oats the proportion of carbohydrates is approximately 57 percent, of fat 5 percent, and of protein 12 percent. These constituents together accordingly yield (57 x 4.1) + (5 x 9.1) + (12 x 4.1), or about 328 calories. An immense number of foods and/or feeds have been analyzed in this manner for their "calorific power." For human needs, tables giving the numbers of calories in common articles of diet are given in any of numerous textbooks on nutrition.** For horses, feed analyses are presented at length by Morrison (1962), Ensminger (1969), Crampton and Harris (1969), in the booklet issued by the National Academy of Sciences (1973), and in a number of other sources, although the foregoing should prove sufficient for reference. It should be added, however, that in

---

* Ibid., p. 134.
** In an average-sized man (69 inches, 155 pounds) at rest, the expected daily need for (or expenditure of) calories is about 1100 per square meter of body surface; and as the surface area in a man of this size is about an even 2 square meters, the number of calories is 2 x 1100, or 2200. In a 1000-pound horse, which because of its relatively large trunk and small limbs in comparison with those of a man has only about 2/3 the body surface of a man in proportion to its weight, the number of calories per square meter of surface is correspondingly higher, about 1600. This, multiplied by the horse's body surface of about 3.9 square meters, gives the "fuel" need as 6260 calories, or 6.26 therms (see Table 22). And, of course, in both man and horse, as the amount of physical activity increases, so also does the need for (or expenditure of) calories.

such analyses differing units of measurement are used according to whether the author favors the Digestible Energy (DE) system or the Total Digestible Nutrients (TDN) system. In our present discussion, which is not proposed to be encyclopedic, Table 23, expressed in the TDN system, gives nutritional analyses of the principal feeds for horses used in the United States today.

17
Daily Requirements
of Nutrients

IN TABLES 22 AND 23, NUTRIENT REQUIREMENTS ARE LISTED IN FOUR CATE-gories according to the degree of physical activity (work) performed, and also in mares during the last quarter of pregnancy and at the peak of lactation, respectively.

Definitions of some of the terms commonly used in connection with animal nutrition are as follows:

Nutrient (N). Any food (or feed) constituent that aids in the support of life. Protein, carbohydrates, fat, minerals, and vitamins are differing forms of nutrients. Air and water, of course, are equally essential.

Digestible Nutrient (DN). That portion, or percentage, of a nutrient which is digestible. The term is generally applied only to the tissue-building or energy-producing nutrients: protein, carbohydrates, and fat.

Total Digestible Nutrients (TDN). The sum of the digestible protein, fiber, nitrogen-free extract, and fat x 2.25.

Digestible Energy (DE). The gross energy of the food intake minus the energy content of the undigested food and the metabolic (body) fraction of the feces. [Where the units of nutrients are expressed in the DE system, they may be converted into TDN units (as used herein) by the formula given in the Appendix.]

Ration. The feed given to an animal during a day (24 hours), whether the feed be given at one time or in portions at several different times.

> *Balanced Ration.* A ration that furnishes the several nutrients in an amount and proportion that will properly nourish a given animal for 24 hours. (Later in this chapter the procedure for computing a balanced ration is described.)

With reference to Tables 22-23, it should be emphasized that even the plus-or-minus of 10 percent generally allowed in nutrient requirements is only an arbitrary estimation. Some horses may require even more, and others even less, than this 10 percent deviation provides for. Competing racehorses, for example, may need amounts of feed, particularly grain or other concentrates, far in excess of those listed for their body weight, even under the condition of hard work. Conversely, there are other horses which seem to do well on quantities of feed noticeably under the average requirements for mature horses of their size. Hence, while Tables 22-23, and the formulas from which it was prepared, provide a reliable basis for estimating *typical or optimum* energy requirements, the amounts prescribed should be modified wherever necessary to conform with the indicated needs of the individual animal.

Although, for consistency, the nutrient requirements listed in Tables 22-25 have been calculated to two decimal places, for all practical purposes one decimal place is sufficient.

However, for the calcium and phosphorus prescribed, which are in much smaller quantities, the listed figures should be taken as they are (that is, to two decimal places for grams and three decimal places for pounds). The Vitamin A requirements are sufficiently accurate as expressed (that is, in whole numbers).

For *maintenance* (mature horses in a state of rest), the daily requirement of dry matter is herein assumed as 1.000. In ratio to this basic figure, the requirement for horses performing light work is 1.300; moderate (medium) work, 1.500; and hard or heavy work, 1.750. In mares during the last quarter of pregnancy the ratio is 1.360, and in mares at the peak of lactation (nursing foals), 1.970.

The recommended amounts of calcium and phosphorus are the same irrespective of the amount of work performed, except in pregnant and in lactating mares, where larger amounts are indicated, particularly in the mares nursing foals. In the first four categories of idle or working horses, the recommended amounts of phosphorus are all 1.1 x the amounts of calcium. In pregnant mares, however, the amounts of each mineral are the same; while in lactating mares, al-

TABLE 22

Daily Requirements of Nutrients for Mature Horses, Ponies, Donkeys, and Mules.
(For intermediate body weights, interpolate the figures listed)

Status of Activity	Body Weight		Dry* Matter,	Digestible* Protein,	Total* Digestible Nutrients,	Calcium†		Phosphorus†		Vitamin A† (carotene),	Net* Energy,
	lbs.	Kg.	lbs.	lbs.	lbs.	grams	lb.	grams	lb.	Mg.	therms
Idle (Relative dry feed = 1.000)	200	91	3.38	0.20	2.34	4.13	.009	4.54	.010	10	1.87
	400	181	5.68	0.33	3.94	6.94	.015	7.63	.017	20	3.15
	600	272	7.71	0.45	5.34	9.41	.021	10.35	.023	30	4.27
	800	363	9.56	0.56	6.63	11.68	.026	12.85	.028	40	5.29
	1000	454	11.30	0.66	7.83	13.80	.030	15.18	.033	50	6.26
	1200	544	12.96	0.76	8.98	15.83	.035	17.41	.038	60	7.18
	1400	635	14.55	0.86	10.09	17.77	.040	19.55	0.43	70	8.06
	1600	726	16.08	0.95	11.15	19.65	.044	21.62	.048	80	8.91
	1800	816	17.56	1.04	12.17	21.44	.048	23.58	.052	90	9.73
	2000	907	19.01	1.12	13.18	23.21	.051	25.53	.056	100	10.53
Light Work: 1–3 hrs. per day of riding or driving (Relative dry feed = 1.300)	200	91	4.40	0.24	3.14	4.13	.009	4.54	.010	10	2.64
	400	181	7.39	0.39	5.30	6.94	.015	7.63	.017	20	4.44
	600	272	10.02	0.54	7.18	9.41	.021	10.35	.023	30	6.02
	800	363	12.43	0.68	8.91	11.68	.026	12.85	.028	40	7.46
	1000	454	14.69	0.81	10.52	13.80	.030	15.18	.033	50	8.83
	1200	544	16.85	0.93	12.07	15.83	.035	17.41	.038	60	10.13
	1400	635	18.92	1.05	13.56	17.77	.040	19.55	.043	70	11.37
	1600	726	20.90	1.16	14.98	19.65	.044	21.62	.048	80	12.57
	1800	816	22.83	1.27	16.35	21.44	.048	23.58	.052	90	13.73
	2000	907	24.71	1.37	17.71	23.21	.051	25.53	.056	100	14.86
Moderate Work: 3–5 hrs. per day of riding or driving (Relative dry feed = 1.500)	200	91	5.07	0.27	3.69	4.13	.009	4.54	.010	10	3.15
	400	181	8.52	0.45	6.20	6.94	.015	7.63	.017	20	5.31
	600	272	11.57	0.62	8.41	9.41	.021	10.35	.023	30	7.19
	800	363	14.34	0.77	11.03	11.68	.026	12.85	.028	40	8.91
	1000	454	16.95	0.91	12.32	13.80	.030	15.18	.033	50	10.55
	1200	544	19.44	1.04	14.13	15.83	.035	17.41	.038	60	12.10
	1400	635	21.83	1.17	15.87	17.77	.040	19.55	.043	70	13.58
	1600	726	24.12	1.30	17.54	19.65	.044	21.62	.048	80	15.01
	1800	816	26.34	1. 42	19.15	21.44	.048	23.58	.052	90	16.39
	2000	907	28.52	1.54	20.73	23.21	.051	25.53	.056	100	17.74

* In these items a range of plus-or-minus 10 percent is allowed from the average amount listed for *mature* animals.

† The specified amounts of calcium, phosphorus, and usually vitamin A at a given body-weight are constant regardless of the amount of work performed, except for pregnant and lactating mares.

In all the feed requirements except vitamin A, and in Net Energy, the amounts are in ratio to Metabolic Body Size (see Table 21). Vitamin A, if measured in U.S.P. units, is 1000 times the quantity in milligrams, as listed in this table. For quantity formulas, see Appendix.

TABLE 23
Daily Requirements of Nutrients . . . (continued).

Daily requirements per animal

Status of Activity	Body Weight lbs.	Kg.	Dry* Matter, lbs.	Digestible* Protein, lbs.	Total* Digestible Nutrients, lbs.	Calcium† grams	lb.	Phosphorus† grams	lb.	Vitamin A† (carotene), Mg.	Net* Energy, therms
Heavy Work: 5–8 hrs. or more daily of riding or driving (Relative dry feed = 1.750)	200	91	5.91	0.31	4.36	4.13	.009	4.54	.010	10	3.79
	400	181	9.94	0.52	7.33	6.94	.015	7.63	.017	20	6.38
	600	272	13.48	0.71	9.94	9.41	.021	10.35	.023	30	8.65
	800	363	16.72	0.88	12.33	11.68	.026	12.85	.028	40	10.72
	1000	454	19.77	1.04	14.57	13.80	.030	15.18	.033	50	12.69
	1200	544	22.68	1.19	16.71	15.83	.035	17.41	.038	60	14.56
	1400	635	25.46	1.34	18.77	17.77	.040	19.55	.043	70	16.39
	1600	726	28.13	1.48	20.73	19.65	.044	21.62	.048	80	18.06
	1800	816	30.72	1.62	22.64	21.44	.048	23.58	.052	90	19.72
	2000	907	33.25	1.75	24.51	23.21	.051	25.53	.056	100	21.34
Mares, last quarter of pregnancy, Light Work (Relative dry feed = 1.360)	200	91	4.60	0.27	3.16	4.66	.010	4.66	.010	12	2.65
	400	181	7.73	0.45	5.32	7.84	.017	7.84	.017	24	4.47
	600	272	10.48	0.61	7.20	10.63	.023	10.63	.023	36	6.05
	800	363	13.00	0.76	8.94	13.19	.029	13.19	.029	48	7.50
	1000	454	15.37	0.90	10.56	15.59	.034	15.59	.034	60	8.87
	1200	544	17.63	1.03	12.11	17.88	.039	17.88	.039	72	10.17
	1400	635	19.80	1.16	13.60	20.08	.044	20.08	.044	84	11.42
	1600	726	21.87	1.28	15.03	22.18	.049	22.18	.049	96	12.62
	1800	816	23.88	1.40	16.41	24.23	.054	24.23	.054	108	13.78
	2000	907	25.85	1.51	17.76	26.22	.058	26.22	.058	120	14.92
Mares nursing foals, Light Work (Relative dry feed = 1.970)	200	91	6.66	0.60	5.22	7.30	.016	6.64	.014	12	4.54
	400	181	11.20	1.01	8.78	12.29	.027	11.18	.025	24	7.63
	600	272	15.18	1.36	11.90	16.65	.037	15.15	.033	36	10.35
	800	363	18.83	1.69	14.76	20.67	.046	18.79	.041	48	12.83
	1000	454	22.26	200	17.45	24.43	.054	22.20	.049	60	15.17
	1200	544	25.53	2.29	20.01	28.02	.062	25.48	.056	72	17.40
	1400	635	28.67	2.58	22.47	31.47	.069	28.61	.063	84	19.54
	1600	726	31.67	2.85	24.83	34.76	.076	31.60	.070	96	21.59
	1800	816	34.59	3.11	27.12	37.96	.083	34.52	.076	108	23.57
	2000	907	37.44	3.36	29.35	41.07	.090	37.35	.082	120	25.52

*† See footnotes under Table 22.

TABLE 24
Daily Requirements of Nutrients for Growing Foals
(For intermediate body weights, interpolate the figures listed)

Mature Body Weight	Body Weight lbs.	Body Weight Kg.	Dry Matter, lbs.	Digestible Protein lbs.	Total Digestible Nutrients, lbs.	Calcium grams	Calcium lb.	Phosphorus grams	Phosphorus lb.	Vitamin A (carotene) Mg.	Net Energy, therms
200 lbs. (91 kg.)	50	23	2.97	.45	1.86	9.4	.021	5.9	.013	3	1.58
	100	45	4.25	.38	2.66	8.1	.018	5.3	.012	6	2.26
	150	68	5.24	.33	3.28	6.8	.015	4.7	.010	9	2.79
400 lbs. (181 kg.)	100	45	4.46	.64	2.79	15.8	.035	10.1	.022	6	2.37
	150	68	5.49	.59	3.43	14.6	.032	9.5	.021	9	2.92
	200	91	6.39	.55	3.99	13.5	.030	8.9	.020	12	3.39
	250	114	7.19	.51	4.49	12.4	.027	8.3	.018	15	3.82
	300	136	7.90	.47	4.94	11.3	.025	7.8	.017	18	4.20
	350	159	8.56	.43	5.35	10.2	.022	7.3	.016	21	4.55
600 lbs. (272 kg.)	100	45	4.68	.85	2.93	22.6	.050	14.2	.031	6	2.49
	150	68	5.78	.79	3.61	21.5	.047	13.7	.030	9	3.07
	200	91	6.71	.74	4.19	20.5	.044	13.1	.029	12	3.56
	300	136	8.29	.67	5.18	18.4	.040	12.1	.027	18	4.40
	400	181	9.62	.61	6.01	16.4	.036	11.1	.024	24	5.11
	500	227	10.82	.55	6.76	14.4	.032	10.1	.022	30	5.75
800 lbs. (363 kg.)	200	91	7.05	.93	4.41	26.7	.059	17.0	.037	12	3.75
	300	136	8.71	.85	5.44	24.7	.054	16.0	.035	18	4.62
	400	181	10.11	.78	6.32	22.8	.050	15.0	.033	24	5.37
	500	227	11.36	.72	7.10	20.9	.046	14.0	.031	30	6.04
	600	272	12.49	.67	7.81	19.0	.042	13.1	.029	36	6.62
	700	318	13.53	.62	8.46	17.1	.038	12.2	.027	42	7.19
1000 lbs. (454 kg.)	200	91	7.40	1.08	4.63	32.4	.071	20.5	.045	12	3.94
	300	136	9.15	1.00	5.72	30.6	.067	19.6	.043	18	5.03
	400	181	10.62	.94	6.64	28.8	.063	18.7	.041	24	5.64
	500	227	11.92	.88	7.45	27.0	.059	17.8	.039	30	6.33
	600	272	13.12	.82	8.20	25.2	.055	16.9	.037	36	6.97
	700	318	14.21	.77	8.88	23.4	.051	16.0	.035	42	7.54
	800	363	15.22	.73	9.51	21.6	.047	15.1	.033	48	8.08
1200 lbs. (544 kg.)	200	91	7.77	1.22	4.86	37.8	.083	23.8	.052	12	4.13
	300	136	9.61	1.14	6.00	36.1	.080	22.7	.049	18	5.10
	400	181	11.15	1.08	6.97	34.4	.076	21.6	.047	24	5.92
	500	227	12.52	1.01	7.83	32.7	.072	20.4	.044	30	6.66
	600	272	13.77	.97	8.56	31.0	.068	19.3	.042	36	7.28
	800	363	15.98	.87	9.99	27.5	.060	18.1	.039	48	8.49
	1000	454	17.95	.78	11.22	24.1	.053	17.0	.037	60	9.54

TABLE 25
Daily Requirements of Nutrients for Growing Foals (continued)

Mature Body Weight	Body Weight lbs.	Body Weight Kg.	Dry Matter, lbs.	Digestible Protein, lbs.	Total Digestible Nutrients, lbs.	Calcium grams	Calcium lb.	Phosphorus grams	Phosphorus lb.	Vitamin A (carotene) Mg.	Net Energy, therms
1400 lbs. (635 kg.)	200	91	8.16	1.35	5.10	43.0	.095	27.0	.060	12	4.33
	300	136	10.09	1.27	6.31	41.3	.091	26.2	.058	18	5.36
	400	181	11.70	1.20	7.31	39.7	.087	25.4	.056	24	6.21
	600	272	14.45	1.09	9.03	36.4	.080	23.7	.052	36	7.58
	800	363	16.77	.99	10.44	33.1	.073	22.1	.048	48	8.87
	1000	454	18.84	.91	11.77	29.9	.066	20.5	.044	60	10.00
	1200	544	20.70	.83	12.94	26.7	.059	18.9	.041	72	11.00
1600 lbs. (726 kg.)	200	91	8.57	1.47	5.35	48.0	1.06	30.1	.066	12	4.55
	400	181	12.29	1.32	7.68	44.8	0.99	28.5	.062	24	6.53
	600	272	15.17	1.20	9.48	41.7	0.92	26.9	.059	36	8.06
	800	363	17.61	1.11	11.01	38.5	0.85	25.3	.055	48	9.36
	1000	454	19.78	1.02	12.36	35.3	0.78	23.7	.052	60	10.51
	1200	544	21.74	.96	13.59	32.1	0.71	22.1	.048	72	11.55
	1400	635	23.56	.88	14.72	29.0	0.64	20.6	.045	84	12.51
1800 lbs. (816 kg.)	200	91	8.99	1.58	5.62	52.8	1.16	33.1	.073	12	4.78
	400	181	12.90	1.43	8.06	49.6	1.09	31.5	.069	24	6.85
	600	272	15.93	1.32	9.96	46.4	1.02	29.9	.066	36	8.46
	800	363	18.49	1.23	11.43	43.3	0.95	28.4	.063	48	10.04
	1000	454	20.77	1.15	12.98	40.2	0.88	26.8	.059	60	11.03
	1200	544	22.83	1.07	14.27	37.1	0.81	25.3	.056	72	12.13
	1400	635	24.74	1.00	15.46	34.1	0.74	23.7	.052	84	13.14
	1600	726	26.52	.93	16.57	31.1	0.68	22.2	.049	96	14.08
2000 lbs. (907 kg.)	200	91	9.44	1.68	5.90	57.5	1.27	35.8	.079	12	5.01
	400	181	13.54	1.54	8.46	54.4	1.20	34.2	.075	24	7.19
	600	272	16.73	1.42	10.46	51.3	1.13	32.7	.072	36	8.89
	800	363	19.41	1.33	12.13	48.2	1.06	31.2	.068	48	10.31
	1000	454	21.81	1.25	13.63	45.1	0.99	29.7	.065	60	11.58
	1200	544	23.97	1.17	14.98	42.0	0.92	28.2	.062	72	12.73
	1400	635	25.98	1.10	16.24	39.0	0.85	26.7	.058	84	13.80
	1600	726	27.85	1.03	17.40	36.0	0.78	25.2	.055	96	14.79
	1800	816	29.58	.97	18.49	33.0	0.72	23.7	.052	108	15.72

though both minerals are increased, the amounts of phosphorus are only .909 x those of calcium.

It may be added that the body weights listed in Tables 22-23 are as low as 200 pounds (91 kg) in order to accommodate very small although mature equines such as Shetland ponies, burros, and "miniature" horses and mules. On the other hand, the weights range upward to 2000 pounds (907 kg) so as to include the needs of draft horses weighing that much. It is assumed, on the basis of available information, that for a given body weight and amount of work performed, the nutritional requirements of mature horses, ponies, donkeys, and mules are identical. For body weights less than 200 pounds or more than 2000 pounds, the nutritional (metabolic) needs may be determined by relating the body weight in pounds$^{0.75}$ to 1000 pounds$^{0.75}$, then applying the derived ratio to the formulas listed in the Appendix. For example, in a Shetland pony weighing 183 pounds the procedure would be as follows: $183^{0.75} = 49.76$; $1000^{0.75} = 177.83$; $49.76/177.83 = .280$. The nutrient needs of the 183-pound pony are therefore seen to be 28 percent of those of a horse weighing 1000 pounds, provided, of course, that the daily amount of work performed (hours ridden or driven) is comparable in the two animals. (If one does not have a convenient table of logarithms from which to calculate the value of a given body weight raised to the $3/4$ power, some of the more versatile electronic pocket calculators now on the market enable such calculations to be made with great ease and speed. If much estimating has to be done, such a calculator will soon make up for its cost.)

While our tables closely parallel those of Morrison, departures are made wherever the majority of specialists in horse nutrition agree on other quantities. And in none of the tables presented by other authors are the quantities of nutrients needed at various body weights and work conditions consistently correlated with Metabolic Body Size, as are the quantities listed here in Tables 22-25. Indeed, in places, some tables present markedly erratic figures, which could not be justified by any of the demands of normal growth. Probably the most variable of these recommended quantities are those of calcium and phosphorus, which according to some authors should be only *half* the amounts listed by others. The greatest discrepancy appears to be in connection with growing foals, where at weaning time the prescribed amounts of calcium and phosphorus may be as much as 50 percent higher than at the age of three months, even though otherwise there is a steady *diminution* of the mineral requirements in relation to the age of the foal. Such recommendations are contrary to the process of normal equine growth, which on the average (the status on which all nutrition must be based) is one of steady *deceleration* from birth to maturity. The healthy growth of the foals of zebras and other wild equids living in a state of nature should demonstrate that no radical changes in nutrition are needed at certain chronological stages. Of course, if a particular young foal should show a subnormal or deficient status of skeletal development. obviously more minerals in the animal's feed should be added. But we are here speaking of basic procedures, not special cases.

The quantities of *dry matter* recommended in Tables 22-23 range from 70 to 80 percent of the total amounts (by weight) of feed to be given daily. For example, a 1200-pound horse at light work should have approximately 16.85 X 1⅓ pounds of feed a day, or about 22 to 24 pounds. It should be noted, as per the nutrients in Tables 22-23 followed by an asterisk, that these items are subject to a plus-or-minus range of 10 percent. This amount of deviation is allowed to accommodate nonconforming cases in which the animal obviously needs more or less feed than the average amounts listed for its body weight and status of activity.

The quantities of *digestible protein* (DP) herein recommended for mature horses vary inversely in relation to those listed for dry matter. For horses at rest, the recommended quantities of DP (by weight) are about 6 percent of those listed for dry matter. This ratio reduces to about 5½ percent in horses at light work and to 5¼ percent in horses at hard work. In mares in the last quarter of pregnancy (during which stage the fetus is growing most rapidly) the ratio is just under 6 percent; but in lactating mares it is increased to 9 percent, or about half-again the latter ratio. These proportions are practically identical with those indicated in the Morrison feeding standards, and similarly allow a plus-or-minus deviation of 10 percent in the case of horses whose nutritional requirements vary from the average. For comparison, the subcommittee on horse nutrition of the National Academy of Sciences (NAS) recommends that the digestible protein be a constant 5.3 percent of the dry matter, both in horses at rest and at work. However, since their recommendation as to dry matter is somewhat higher than mine, the quantity of DP under both systems works out to be about the same. But there is a difference in the case of pregnant mares, where the NAS recommends 6.9 percent as compared with my 5.86 percent. In lactating mares, the NAS listing diminishes from 8.7 percent at a body weight of 200 kg to 8.0 percent at 600 kg, whereas the recommendation herein is a constant 9.0 percent at all body weights. Both the NAS and the Morrison feeding standards, as well as those presented here, list smaller amounts of protein than were prescribed by earlier authorities, since various experiments have shown that the smaller amounts, both in working as well as resting horses, are adequate. Too, by obviating the need for protein-rich feeds, they save the horse owner unnecessary expense. However, neither is it advisable to feed rations that are too low in protein.

The quantities of *total digestible nutrients* (TDN) recommended herein are roughly 70 percent of those listed for dry matter. The actual range is from 69.3 percent for horses at rest to 73.7 percent for horses at hard work. In pregnant mares the ratio is 68.7 percent, while in lactating mares it is increased to 78.4 percent. But these theoretical average figures are elastic, and for all practical purposes can be converted to the nearest whole numbers. For those who use the *digestible energy* (DE) system for calculating feed rations, the DE in Megacalories (Mcal) in horses at rest is equivalent to .515 x TDN, lbs.; in horses at light work, .532; at medium work, .540; and at hard work, .548. In pregnant mares the conversion ratio (DE/TDN) is .510; and in lactating mares, .583. For example, the daily TDN requirements in a 1000-pound horse at light work is 10.52 lbs. (per Tables 22-23). Hence the corresponding requirement in DE is .532 x 10.52, or 5.60 Mcal. Since Table 26 lists both the TDN and the DE percentages, a balanced feed ration may readily be calculated by using either of the two systems.

The recommended amounts of vitamin A (carotene, or vitamin A as derived from vegetable sources) listed in Tables 22-23 are sufficiently exact to be expressed as a constant percentage of body weight *per se*, rather than of Metabolic Body Size. For all classes of horses either idle or at work, the prescribed daily amount of vitamin A in Megagrams is 5 percent ($\frac{1}{20}$) of the body weight; and for pregnant or nursing mares, 6 percent.

In calculating the foregoing nutritional requirements in *growing foals*, procedures similar to those used for mature horses have been adopted, with, of course, proper allowances being made for the requirements of growth in addition to those for maintenance. One of the main differences in this respect is that in foals the proportion of digestible protein to dry matter may range from 15 to 18 percent, whereas in mature horses it is usually only 5 or 6 percent, except in lactating mares, where it may rise to 9 percent. Similarly, in the re-

TABLE 26
COMPOSITION OF FEEDS

Feedstuff	Total dry matter %	Protein %	Digestible protein %	TDN %	DE* therms/lb.	Calcium %	Phosphorus %	Vitamin A (Carotene) mg/lb.
				Dry Roughages				
Alfalfa hay, all analyses	90.5	15.3	10.9	50.7	1.02	1.47	0.24	8.2
Alfalfa hay, past bloom	90.5	12.9	9.3	47.7	0.96	1.10	0.20	3.3
Alfalfa meal, dehydrated	92.7	17.7	12.4	54.4	1.10	1.60	0.26	42.4
Bromegrass hay, all analyses	88.8	10.4	5.3	49.3	1.00	0.42	0.19	—
Clover hay, alsike, all analyses	88.9	12.1	8.1	53.2	1.07	1.15	0.23	—
Clover hay, Ladino	89.5	18.5	14.2	59.5	1.20	1.53	0.29	—
Clover hay, red, all analyses	88.3	12.0	7.2	51.8	1.05	1.28	0.20	7.3
Clover & timothy hay, 30 to 50% clover	88.1	8.6	4.7	51.0	1.03	0.69	0.16	—
Orchard grass hay, good	88.7	8.1	4.2	49.7	1.00	0.27	0.18	—
Reed canary grass hay	91.1	7.7	4.9	45.1	0.91	0.33	0.16	—
Sudan grass hay, all analyses	89.4	8.8	4.3	48.6	0.98	0.36	0.27	—
Timothy hay, all analyses	89.0	6.6	3.0	49.1	0.99	0.35	0.14	4.4
Timothy hay, full bloom	89.0	6.4	3.2	51.1	1.03	—	0.20	4.2
Timothy and clover hay, ¼ clover	88.8	7.9	4.0	49.8	1.01	0.58	0.15	—
Wheat straw	92.6	3.9	0.3	40.6	0.82	0.15	0.07	—
				Silages				
Carrots, roots	11.9	1.2	0.9	10.3	0.21	0.05	0.04	—
				Concentrates				
Barley, excluding Pacific Coast	89.4	12.7	10.0	77.7	1.57	0.06	0.40	—
Beet, pulp, dried	90.8	9.1	4.3	68.2	1.38	0.68	0.10	—
Bone meal, steamed	95.2	12.1	—	—	—	28.98	13.59	—
Corn, yellow dent, #2	85.0	8.7	6.7	80.1	1.62	0.02	0.27	1.3
Cottonseed oil meal, solvent	91.4	41.6	34.5	66.1	1.34	0.15	1.10	—
Linseed oil meal, solvent	90.9	35.1	29.5	71.0	1.43	0.40	0.83	—
Molasses, beet	76.0	6.7	3.5	59.6	1.20	0.16	0.03	—
Molasses, cane	74.5	7.0	3.2	54.9	1.11	0.89	0.08	—
Oats, excluding Pacific Coast	90.2	12.0	9.4	70.1	1.42	0.09	0.33	—
Soybean oil meal, solvent	89.3	45.8	42.1	77.2	1.56	0.32	0.67	—
Wheat, hard, winter	89.4	13.5	11.3	79.6	1.61	0.05	0.42	—
Wheat, soft, winter	89.2	10.2	8.6	80.1	1.62	—	0.29	—
Wheat bran	89.1	16.0	13.0	65.9	1.33	0.14	1.17	1.2
Yeast, brewers dried	93.4	44.6	38.4	72.4	1.46	0.13	1.43	—

* DE (digestible energy) may be converted to metabolizable energy by multiplying by 82 per cent.

(This convenient table is from the Cornell University Equine Research Program: HORSE FEEDING, by Drs. H. F. Hintz and H. F. Schryver.)

quirements of calcium and phosphorus, those of foals may be from 2 to 3 times as great (in proportion to dry matter) as in mature horses. Of vitamin A, however, the need in foals is generally considered to be about 6 percent of the body weight, which is the same proportion as that recommended for pregnant and lactating mares; while for horses in the various work classes the ratio is 5 percent.

Net energy in the needs of foals is similar to that in mature horses, averaging 85 percent of the TDN, while among the mature horses it ranges according to the amount of work performed from 80 to 87 percent. It may be added, however, that the need for this element (net energy) is stated quite variously by different authors. Brody, et al (Bull. 368, pp. 6–11, 1943), for example, presents equations that make the daily caloric (= therm) requirement up to nearly *twice* what Morrison lists. Hence, all the specifications of nutritional requirements presented by any author can at best be only guides or approximations, the optimum amounts of which may in each case be established only by checking the growth and general physical condition of the foal against the average or typical gains made during the age interval in question. Figure 4 and Table 8 should be helpful guides in this procedure.

It is now opportune to list some appropriate daily rations based on the nutritional recommendations made in Tables 22-23.

Sample Daily Rations for Foals, Yearlings, and Two-Year-Olds

Shortly after the foal is born, it should take a good draught of colostrum, or first milk of the mare. This milk possesses purgative properties that help to discharge from the alimentary tract the waste products accumulated therein during fetal life. If this cleaning-out is not accomplished, a small dose of castor oil should be given the foal. If he is obtaining an oversupply of milk, he may have an attack of diarrhea. In that case some of the mare's milk should be withdrawn. The foal's diarrhea should be checked at once through giving either rice meal gruel, parched flour, or boiled milk and whites of raw eggs.

A properly fed foal will attain from 50 to 60 percent of its potential adult weight during its first year (see Table 6). After that, a marked increase in the amount of feed is necessary to continue the weight gain. The foal will subsist on its mother's milk for the first three to four weeks, after which it will show interest in grass, hay, and grain, and will start to take solid food from the mother's trough.

Table 27

Sample Daily Rations for a Mature Horse (or Mule) Weighing 1000 lbs.*

(Poundages listed equal from 1.3 to 1.4 x Dry Matter, depending upon differences
in Energy content. See Tables 21a and 21b.)

Recommended Total Rations, lbs.	Choices of Feedstuffs
When idle 14.7 (13.2 – 16.2)	1. Legume hay, 14.5 lbs. 2. Grass hay, 14 lbs., and a high-protein concentrate (such as 3. Legume hay, 7.4 lbs.; grass hay, 7.5 lbs. linseed meal) 0.60 lb. 4. Oat or barley straw, 4.7 lbs.; legume hay, 9.4 lbs. 5. Corn or sorghum stover, 8.6 lbs.; legume hay, 6.4 lbs. 6. Corn or sorghum silage, 11.6 lbs.; oat or barley straw, 4.6 lbs.
At light work 19.1 (17.2 – 21.0)	1. Grass hay, 14 lbs.; rolled oats, 5.2 lbs. 2. Grass hay, 14 lbs.; cracked corn, 3.9 lbs.; linseed meal or other high-protein supplement, 0.43 lb. 3. Legume hay, 14 lbs.; cracked corn, 3.5 lbs.; molasses, 1 lb. 4. Legume hay, 7 lbs.; grass hay, 7 lbs.; corn, 3.9 lbs. 5. Shredded corn fodder, 7 lbs.; legume hay, 7 lbs.; rolled oats, 4.3 lbs. 6. Oat or barley straw, chopped, 5.2 lbs.; legume hay, 8.7 lbs.; rolled oats, 5.2 lbs.
At medium work 22.0 (19.8 – 24.2)	1. Grass hay, 12.2 lbs.; rolled oats, 9.6 lbs. 2. Grass hay, 12.2 lbs.; cracked corn, 7.8 lbs.; linseed meal or other high-protein supplement, 0.65 lb. 3. Legume hay, 12.2 lbs.; cracked corn, 7.8 lbs.; molasses, ½ lb. 4. Shredded corn fodder, 5.2 lbs.; legume hay, 5.2 lbs.; rolled oats, 13 lbs. 5. Legume hay, 5.2 lbs.; grass hay, 5.2 lbs.; cracked corn, 12.2 lbs. 6. Oat or barley straw, chopped, 3.5 lbs.; legume hay, 7 lbs.; rolled oats, 14 lbs.
At hard work 25.6 (23.0 – 28.2)	1. Grass hay, 10.4 lbs.; rolled oats, 14 lbs.; molasses, 1 lb. 2. Grass hay, 10.4 lbs.; cracked corn, 11.3 lbs.; linseed meal or other high-protein supplement, 7/8 lb. 3. Legume hay, 10.4 lbs.; cracked corn, 11.7 lbs. 4. Legume hay, 5.2 lbs.; grass hay, 5.2 lbs.; cracked corn, 12.2 lbs. 5. Shredded corn fodder, 5.2 lbs.; legume hay, 5.2 lbs.; rolled oats, 13 lbs. 6. Oat or barley straw, chopped, 4.3 lbs.; legume hay, 7 lbs.; rolled oats, 14 lbs.
Lactating mares, not at work 20.0 (18.0 – 22.0)	1. Alfalfa, soybean, or cowpea hay, 14 lbs.; corn or other grain, 5.2 lbs. 2. Red clover hay, 14 lbs.; rolled oats or ground barley, 2.6 lbs.; cracked corn, 2.6 lbs. 3. Mixed clover-and-timothy hay (containing 30 percent or more clover), 14 lbs.; rolled oats, 5.2 lbs. 4. Timothy or other grass hay, 14 lbs.; rolled oats, 2.6 lbs.; bran, 2.6 lbs.; linseed meal or other high-protein supplement, 7/8 lb.
Mares in last quarter of pregnancy, not at work 21.0 (20.0 – 22.0)	1. Grass-alfalfa hay, 15 lbs.; rolled oats, 3 lbs.; cracked corn, 21 bs.; wheat bran, 1 lb.; molasses, 1 lb. 2. Timothy hay, 7 lbs.; alfalfa hay, 7 lbs.; rolled oats or ground barley, 3 lbs.; shelled corn, 3 lbs. 3. Clover hay, 15 lbs.; rolled barley, 4 lbs.; soybeans or cowpeas, ground, 1 lb.; molasses, ½ lb.
Breeding Stallion: light breeding program	Timothy-clover hay, 10 lbs.; rolled oats, 4 lbs.; cracked corn, 2 lbs.; wheat bran, 1 lb.; molasses, 1 lb.; soybean oil meal, ½ lb.
ditto, medium breeding program	Brome-alfalfa hay, 12 lbs.; rolled oats, 7 lbs.; wheat bran, 2 lbs.; molasses, 1 lb.
ditto, heavy breeding program	Brome-alfalfa hay, 12 lbs.; rolled oats, 9 lbs.; wheat bran, 3 lbs.; molasses, 1 lb.; soybean oil meal, 1 lb.

* For body weights other than 1000 pounds, multiply the quantities (weights) of feeds in the above list by the appropriate factor in the second column of Table 21.

This should be allowed before letting the mare eat. When the foal reaches the age of 5 or 6 weeks, it is advisable to make a creep feeder* available to it, unless rich pasturage is at hand. The creep type of feeding is adequate until weaning time (5 to 6 months of age). The creep feeding schedule also makes the weaning separation less difficult for the foal. The following daily rations are suggested for foals from the creep stage to two years of age (1000 lbs. mature weight).

Daily Creep Rations (of 100-pound trough mixtures)	1. Rolled oats, 60 lbs.; calf manna, 40 lbs. 2. Rolled oats, 50 lbs.; wheat bran, 40 lbs.; linseed oil meal, 10 lbs. 3. Rolled oats, 30 lbs.; barley, 30 lbs.; wheat bran, 30 lbs.; linseed oil meal, 10 lbs. 4. Rolled oats, 80 lbs.; wheat bran, 20 lbs. 5. Rolled oats, 60 lbs.; cracked corn, 20 lbs.; calf manna, 20 lbs.
Weaning Foals, Daily Rations (5–6 months of age, 350–400 lbs.)	Timothy-alfalfa hay, 6¼ lbs.; Rolled oats, 2¾ lbs.; Ground corn, 1½ lbs.; Wheat bran, 1½ lbs.; Molasses, ½ lb.
Foals, 8 months of age (400–450 lbs.)	Timothy-alfalfa hay, 7¾ lbs.; Rolled oats, 2¾ lbs.; Ground corn, 1½ lbs.; Wheat bran, 1 lb.; Molasses, ¾ lb.; Soybean oil meal, ½ lb.
Yearlings (500–600 lbs.)	Timothy-alfalfa hay, 9 lbs.; Rolled oats, 2 lbs.; Ground corn, 2¾ lbs.; Wheat bran, 1¼ lbs.; Molasses, 1 lb.
Two-Year-Olds (750–850 lbs.)	Grass hay, 12½ lbs.; Rolled oats, 2¾ lbs; Wheat bran, 1 lb.; Molasses, 1 lb.; Soybean oil meal, ½ lb.

---

* A creep is a feeding enclosure for young foals. The fencing is constructed so that the foals may enter, but not the mares, which should be encouraged to stay near the creep by providing them with fresh water and a salt supply. Inside the creep should be the foals' feeding trough, in which at first should be placed only a small amount of feed, the surplus being removed each day and given to the mares. Salt and a suitable mineral mixture should be provided in the creep. (A detailed plan for building a creep feeder may be obtained free of charge from Ralston Purina Co., Checkerboard Square, St. Louis Mo. 63188.)

18
Computing a Balanced Ration

IT HAS BEEN SAID THAT A GOOD HORSEMAN SHOULD KNOW HOW TO balance a ration. A balanced ration is one that supplies the various essential nutrients in such proportion and amount as will properly nourish a given animal for 24 hours. To compute a balanced ration, several steps are necessary. First, it must be known what the nutritional requirements for the particular animal are. To give an example, let us put down these needs for a 1000-pound horse performing light work (i.e., from 1 to 3 hours per day of riding or driving). From Table 23 we derive these figures (the first three items and Net Energy being expressed additionally in their 10 percent plus-or-minus range).

Dry Matter, lbs.	Digestible Protein, lbs.	Total Digestible Nutrients, lbs.	Calcium, lbs.	Phosphorus
14.69* (13.2–16.2)	0.81 (0.7–0.9)	10.52 (9.5–11.6)	.030	.033

The next step is to see if these recommended allowances may be met by selecting (through "trial-and-error") a certain ration; say, in this case, of grass hay and oats. The percentages by which the weights of hay and oats are multiplied are those listed under these feeds in Table 26.

Timothy hay, 14 lbs.	3% of 14 lbs. = 0.42 (together = 0.80 lb.)	49.1% of 14 lbs. = 6.87 (together = 9.67 lbs.)	0.35% = .0490	0.14% = .0196
Oats, 4 lbs.	9.4% of 4 lbs. = 0.38	70.1% of 4 lbs. = 2.80	0.09% = .0036	0.33% = .0132

It is thus seen that this initial estimate is sufficiently close to provide the essential amounts of nutrients. The 14 pounds of timothy hay and 4 pounds of oats together supply .80 pound of Digestible

---

* For all classes of work performed, the average ratio of Dry Matter to Total Daily Feed, as derived from feed recommendations in general, ranges between 70 and 80 percent. Conversely, the daily amount of feed ranges between 1.25 and 1.40 times the weight of Matter listed in Tables 22 and 23.

TABLE 28

*Weights and Measures of Common Feeds**

Feeding stuff	Average weight, lbs.		Quarts
	per quart	per bushel	per pound
Roughages			
Alfalfa meal	0.6	19	1.7
Buckwheat hulls	0.5	16	2.0
Cottonseed hulls	0.3	10	3.3
Oat hulls	0.4	13	2.5
Concentrates			
Barley, whole	1.5	48	0.7
Barley, ground	1.1	35	0.9
Beans, field	1.7	54	0.6
Beet pulp, dried	0.6	19	1.7
Brewers' grains, dried	0.6	19	1.7
Buckwheat ,whole	1.4	45	0.7
Buckwheat feed	0.6	19	1.7
Buckwheat flour	1.6	51	0.6
Buckwheat middlings	0.9	29	1.1
Citrus pulp, ground	1.0	32	1.0
Coconut oil meal	1.5	48	0.7
Corn, dent, whole	1.7	56	0.6
Corn, dent, ground	1.5	48	0.7
Corn-and-cob meal	1.4	45	0.7
Corn bran	0.5	16	2.0
Corn germ meal	1.4	45	0.7
Corn gluten feed	1.3	42	0.8
Corn gluten meal	1.7	54	0.6
Cottonseed, whole	0.8	26	1.3
Cottonseed meal	1.5	48	0.7
Cowpeas	1.7	54	0.6
Distillers' corn grains, dried	0.6	19	1.7
Flaxseed	1.6	51	0.6
Flaxseed screenings	1.1	35	0.9
Hominy feed	1.1	35	0.9
Linseed meal, old process	1.1	35	0.9
Linseed meal, new process	0.9	29	1.1
Malt sprouts	0.6	19	1.7
Millet seed, foxtail	1.6	51	0.6
Molasses, cane	3.0	96	0.3
Molasses feeds	0.8	26	1.3
Oats	1.0	32	1.0
Oats, ground	0.7	22	1.4
Oatmeal, without hulls	1.7	54	0.6
Oat middlings	1.5	48	0.7
Oat mill feed	0.8	26	1.3
Peas, field	2.1	67	0.5
Rice bran	0.8	26	1.3
Rice polish	1.2	38	0.8
Rye, whole	1.7	54	0.6
Rye, ground	1.5	48	0.7
Rye bran	0.8	26	1.3
Rye feed	1.3	42	0.8
Rye middlings	1.6	51	0.6
Soybeans	1.8	58	0.6
Sunflower seed	1.5	48	0.7
Tankage	1.6	51	0.6
Wheat, whole	1.9	60	0.5
Wheat bran	0.5	16	2.0
Wheat mixed feed	0.6	19	1.7
Wheat flour middlings	1.2	38	0.8
Wheat screenings	1.0	16	1.0
Wheat standard middlings	0.8	26	1.3

* Adapted from *Feeds and Feeding*, by F. B. Morrison (22nd ed., 1959 p. 1114).

TABLE 29

Suggested Daily Rations for Horses Performing Specific Kinds of Work

Type of horse and work required	Daily Ration	
	Concentrates	Roughage
Trotters:		
Colt, weaning time	2 pounds oats	Hay, ad libitum
" one year old	4 " "	" " "
" two years old	6 " "	" " "
" " " " (in training)	8 " "	Hay, limited amount
" 3 years old (in training)	8–12 " "	" " "
Adult horse, racing	10–15 " "	" fair amount
Thoroughbreds:		
Adult, racing	15 pounds oats*	6–8 pounds hay
" riding	8 " "	12 " "
Horses variously used:		
Pony (c. 350 pounds)	4 pounds oats	Hay, moderate amount
Hunter, small (c.1100 lbs.)	12 " "	12 pounds hay
" large (c.1300 lbs.)	16 " "	10 " "
Light Carriage (c.1000 lbs.)	10 " "	12 " "
Draft horses:		
At heavy, hard work	13 pounds oats, 6 pounds beans, 3 pounds corn	15 pounds chopped clover hay
Farm horses:		
At light work	6–10 pounds oats	6–9 pounds hay, 3 pounds straw
At medium work	10 " "	10 pounds hay, 3 pounds straw
At heavy work	13 " "	12 pounds hay, 3 pounds straw

---

* In 1973, Secretariat, who was larger than the average three-year-old Thoroughbred, was given 16 pounds of oats a day, along with enough hay to keep him nibbling "almost constantly." Secretariat's evening feed consisted usually of "oats cooked into a mash, plus carrots and some vitamins and minerals, plus some . . . grains coated with molasses to provide the rough equivalent of a candied breakfast cereal." (*Time,* June 11, 1973, p. 87.)

Protein and 9.67 pounds of TDN, while the calcium supplied in these two feeds is well above the required amount and the phosphorus practically identical thereto.

However, the rations for a particular horse under specific conditions of age, size, sex, and work conditions may require correspondingly involved feed computations, to an extent which is beyond the scope of the present book. For horsemen and breeders desirous of having feed analyses in great detail, the specialized works on this subject listed in the Bibliography—especially those by Morrison, Ensminger, and the National Academy of Sciences, respectively—are heartily recommended.

Another method of balancing a ration, by what is known as the "nutritive ratio" of a feed, is described in detail by J. T. Willard in the *Cyclopedia of American Agriculture*, Vol. III (Animals), pp. 103–5 (see Bibliography).

For comparison with the rations recommended by the author, the following earlier (1908) table is presented. It was contributed by Merrit W. Harper in his article, *Feeding the Horse*, in the *Cyclopedia of American Agriculture*, Vol. III (Animals), pp. 103–5. Note that the quantities of oats and hay prescribed by Harper are essentially the same as those listed for the corresponding work classes in Table 27.

19

A Brief Review of Rules and Hints for Feeding and Caring for a Horse

THE RECOMMENDATIONS THAT COULD BE MADE FOR PROPERLY CARING for horses could readily fill a volume. The following observations represent a few of those agreed upon by horsemen generally. They are not necessarily listed in order of importance.

Efficient feeding may save up to 20 percent of the usual feed cost. However, it is wiser to purchase feeds of known high quality than to attempt to save money by using inferior products. Avoid any feeds—whether commercial or home-grown—that contain dust or mold. Certain molds—for example, moldy hay—can cause disease; and special care should be taken not to feed spoiled silage. Horses and sheep are more apt to suffer from spoiled feed than are cattle; and even swine have died from eating moldy corn.

A basic rule of feeding is "little and often." Four feedings a day are better than three, and three better than two. Since a horse has a relatively small stomach, it is essential that he be not overfed; otherwise, colic or some other digestive disorder may result. Horses vary more than other farm animals in individual feed requirements, disposition, and taste. If a horse is tired, he should be given a smaller quantity of food, of an easily digested nature.

Since a horse's digestive system is not adapted for concentrated food, it is necessary that he be fed plenty of hay or other roughage. Except when the horse is at hard work, good pasturage can substitute for hay or other dry fodder. The general rule for feeding *working* horses is to reduce the amount of hay (or other roughage) and increase the amount of grain (or other concentrate) as the work is increased. Thus a 1000-pound horse when idle might be fed 14 pounds of hay and a small (less than a pound) amount of a concentrate, whereas if at hard work the hay could be reduced to 10 pounds while the concentrate (such as oats or barley) could be increased to 13 or 14 pounds.

As mentioned throughout this section, the amount of feed a horse is given should be based on his size and the amount of work he (or she) performs. Thus, at a given body weight, a racehorse in action may require more than *twice* the amount of feed he needs when idle. The feed in such case should be rich in protein, TDN, vitamins, and minerals. Among protein supplements, wheat bran is especially useful because of its bulk and its laxative effect. Other commonly used protein supplements are soybean oil meal, cottonseed meal, and corn gluten feed.

Horses should be given fresh, clean drinking water from one to three times a day. Avoid public feeding and watering facilities. Before running a race, a horse should have no water for three or four hours. Under ordinary conditions a horse should be given water first, then hay or other roughage, then grain or other concentrate. If the horse is hot from exercise, he should be given only small amounts of water until he has cooled down.

As mentioned above, only clean feed of good quality should be given. The most desirable roughages are properly-cured, mold-and-dust-free green hays derived from choice alfalfa and timothy. In summer, good rolled oats should be the mainstay of the feed ration; and in winter a balanced ration of grains, corn, soybean oil meal, minerals and vitamins. The better the *quality* of the feed, the higher the proportion of digestible nutrients.

Horses not at pasture should be fed at approximately the *same hours* each day. A horse should be fed at least an hour or two before work starts, and should *not* be worked on a full stomach. If three meals per day are given, the noon meal should be lightest, the morning meal heavier, and the evening meal nearly or fully equal in quantity to the two earlier feedings combined. Never make *sudden changes* in the kinds and amounts of feed given, nor in the hours at which the feeding is done. On days that working horses are idle, the grain ration should be reduced by ½, and the hay ration increased.

In the northern states, the principal roughage is timothy hay, while in the South this is replaced by various grasses such as Bermuda, Johnson, and/or Dallis. Grass hay has the advantage of being usually free from mold or dust. It also more closely approaches legume hay in its nutritive value for horses than for other kinds of livestock.

The feeding of each individual horse should be considered *separately,* in view of the animal's age, condition, temperament, and amount of work required. All feeding should be done by *weight,* not volume, and as nearly as possible at the *same hours* each day (Table 25 gives the weights of various common feeds in relation to their quantities in quarts and bushels). Any changes in the feed given should be made *gradually.* The body weights of foals should be periodically checked against the growth rates and tables given in Part I of this book.

In order for a horse to derive full benefits from his feed, it is advisable that he be checked at regular intervals (at least twice a year) for *worms* and as to the condition of his *teeth.* If worms are present in the horse's droppings, a veterinarian should be called in to administer liquid deworming medicine directly to the animal's stomach. This is usually performed by "tubing," which consists of passing a flexible tube down the esophagus and to the stomach. At similar intervals the horse's teeth should be examined for decay and for any sharp points or edges, resulting from wear, that should be filed down. If a horse has bad teeth and cannot chew his feed properly and without discomfort, he will be unable to gain weight no matter how much he is fed.

As to the properties or qualities of various feeds, here are some opinions from Captain Horace Hayes, an eminent authority on the care and handling of horses of all breeds throughout the British Empire:

Oats are superior to all other kinds of grain. Good English oats weigh from 37 to 48 pounds a bushel (in southern USA, sometimes as little as 20 pounds). However, there is no feeding advantage in oats weighing over 40 pounds a bushel. Maize is second in value (used in South Africa).

The large percentage of *fat* in oats and maize undoubtedly enhances the value of these grains as foods for horses, a small amount of fat (fuel) being a necessary constituent of food, even in herbivorous animals.

Pigs fed on food rich in nitrogen—i.e., legumes and oily or fatty foods—developed larger muscles and stronger bones than those fed on foods rich in carbohydrates but poor in nitrogen (protein).

Corn is *absolutely necessary* for hard-worked horses.

Vegetable feeders, like the horse, have far less power of absorbing fat from food than carnivora, which in a state of nature consume little or no starch or sugar.

The fact that fat horses can live far longer than thin ones, when both are deprived of all food except water, shows that a food rich in fat-formers supplies much more energy than a food rich in muscle-formers (protein).

Bone is better developed from grass or alfalfa than from corn. Hence the "light bone" of Thoroughbreds. However, large bones and great speed are not compatible. But bone of fairly good size is essential for weight-carrying purposes.

As a concentrate feed for horses, barley is not much inferior to maize. It should be broken and mixed with chopped hay and chopped straw. Barley is the only grain that is generally used for horses in Syria, Egypt, Arabia, Persia, Algeria, and other Eastern countries. [This was written by Hayes in 1905.]

For draft horses at hard work, barley is a food equally as good as oats. A good daily ration for such horses would be: hay 13 lbs., beans 3 lbs., maize 8 lbs., barley *or* oats 8 lbs.

Wheat may be dangerous if pasty (difficult to digest) and not ground coarse enough or mixed with bran.

Linseed is the seed of the flax plant; it contains 37% fat and oil. Consequently it is useful for fattening horses that are "low in flesh." It improves the *coat* markedly.

Beans and peas are useful only as an adjunct to grain rich in starch.

The usual amount of water contained in a horse's feed should be between 20 and 25 percent, by weight. If the food is too watery, digestion is hastened and nutrition diminished. And if the food is too dry, digestion is slowed and constipation encouraged. Dry corn, for example, should be supplemented with fresh green grass or fodder, say one hour before or two hours after the corn.

Nutritional deficiencies in horses are manifested by lowered energy and endurance, emaciation, poor bone development, slow healing, and various other aspects of malnutrition, depending upon which particular element or elements are lacking in the diet. The lack may be either in protein, or in any of the various minerals or vitamins. The age of the horse, its productive level, and the geographic locality where the animal forages are also factors in its nutritional status. Since the deficiencies are usually multiple and often difficult to diagnose, it is generally advisable to summon a veterinarian to examine the afflicted animal and prescribe treatment.

Nutritional diseases in horses may be caused either by improper feeding, working, or watering; a lack of minerals or of vitamins; or following an infection. Some common diseases of this nature are: anemia (nutritional), azoturia ("Monday morning disease"), colic, founder, goiter (iodine deficiency), heaves, periodic ophthalmia ("moonblindness"), rickets (deficiency of calcium, phosphorus, or vitamin D), salt deficiency, urinary calculi or stones (usually only in males), "night-blindness" (vitamin A deficiency), and osteomalacia (softening of bone, caused by lack of vitamin D, calcium, and phosphorus). While some of these afflictions may be overcome by adding to the horse's feed the nutrients, minerals, or vitamins in which the feed is lacking or deficient, it is generally safer to call in a veterinarian to make a reliable diagnosis and prescribe proper treatment.

20
Signs of a Well-fed,
Healthy Horse*

THE OUTWARD MANIFESTATIONS OF A PROPERLY-FED, HEALTHY AND spirited horse are as follows:

1. The horse appears at ease and "unworried" when resting.
2. The horse is alert, "bright-eyed", and attentive.
3. One of the attributes of a healthy, well-conditioned horse is a soft, glossy coat and a pliable, elastic skin. When the hair loses its luster and the skin becomes dry, scurfy and hidebound, there is usually trouble ahead.
4. A healthy horse has a good appetite, and will neigh and paw before being fed and go after its feed with relish.
5. The membranes of a horse's eyes, which can be seen when the lower lid is pulled down, should be pink and moist.
6. A healthy horse should have feces and urine that are normal. The feces should be neither too dry nor too loose; and both the feces and urine should be excreted without straining and be free of blood, mucus, or pus.
7. The normal temperature of a horse, taken rectally, ranges between 99 and 100.8 degrees, averaging 100.5 degrees F. The normal pulse rate ranges between 32 and 44 beats per minute, and the normal breathing rate from 8 to 16 breaths per minute.
8. In addition to the foregoing signs of health in a horse, there are others, such as a sound set of teeth; well-muscled neck, body, and limbs; and strong, well-formed hoofs. However, it should be added that a horse may have "unseen" qualities, such as lung capacity, digestive ability, all-around working capacity, and even intelligence and disposition that cannot be determined solely from physical inspection.

* Based in part on USDA Agriculture Handbook No. 394 (1972), p. 39.

Appendix: Part I

1. Formulas Used in Correlating Various Body Measurements

THE FOLLOWING FORMULAS ARE THOSE WHICH WERE USED FOR CORRE-
lating the principal body measurements introduced in Part I. These
formulas are not presented as final, inflexible laws of growth, but are
merely the best equations that the writer has been able to evolve
from the data at his disposal. Most authors, in making correlations
of this kind, adopt formulas in which logarithms are used. Here, however, the preference has been given to simple, arithmetical equations
which give as good correlations* and are easier for the general reader to understand. Actually, the whole field of equine body measurements is still in need of more extensive exploration. It is hoped that
the following formulas, as determined in this study, will prove of assistance to future students of hippometry.

Body Weight, pounds**

(All breeds except ponies)

Adult males = (.14475 Chest Girth, inches)³
Adult females = (.14341 Chest Girth, inches)³
Male foals, up to 66 inches Chest Girth =
(.1387 Chest Girth, in. + 0.400)³
Female foals, up to 66 inches Chest Girth =
(.1382 Chest Girth, in. + 0.344)³
Newborn males = .0854 Body Weight of Sire, + 13.15 pounds
Newborn females = .0854 Body Weight of Dam, + 13.15 pounds

An exception is in the Shetland, where the ratio is .9629 male
(5.19 : 5.39) and would be expected to be higher. Possibly this ratio is due to the influence of "cold" (draft build) in the Shetland.
The most reliable estimation of body weight is one in which five

---

* These inter-correlations yield measurements generally accurate within ± 1 percent,
and usually within ± ½ percent. In actual practice, no such accuracy is required, as
the chest girth can hardly be measured closer than within one inch and the cannon
girth than within 1/10 inch.
** For estimating the weights of Shetlands, see directions on p. 38.

Chest Girth, inches*

All breeds except ponies

Adult males $= 6.9084 \sqrt[3]{\text{Body Weight}}$

Adult females $= 6.9730 \sqrt[3]{\text{Body Weight}}$

Male foals, up to 870 pounds Body Weight $=$
$$7.210 \sqrt[3]{\text{Body Weight}} - 2.89$$

Female foals, up to 850 pounds Body Weight $=$
$$7.236 \sqrt[3]{\text{Body Weight}} - 2.49$$

Light Horse males, birth to adult $= 57.21 \sqrt{\text{Cannon Girth}} - 92.00$

Light Horse females, birth to adult $= 62.58 \sqrt{\text{Cannon Girth}} - 102.29$

Draft Horse males, birth to adult $= 11.11 \text{ Cannon Girth}, - 28.99$

Draft Horse females, birth to adult $= 12.59 \text{ Cannon Girth}, - 35.85$

Shetland Pony males, birth to adult $= 14.92 \text{ Cannon Girth}, - 31.47$

Shetland Pony females, birth to adult $= 17.04 \text{ Cannon Girth}, - 38.61$

(Fore) Cannon Girth, inches

Light Horse adult males $= .9913 \sqrt[3]{\text{Body Weight}} - 2.26$

Light Horse adult females $= .8545 \sqrt[3]{\text{Body Weight}} - 1.13$

Light Horse males, birth to adult $= .01748 \ (\text{Chest Girth} + 92.00)^2$

Light Horse, females, birth to adult $= .01598 \ (\text{Chest Girth} + 102.29)^2$

Draft Horse males, birth to adult $= .0900 \text{ Chest Girth}, + 2.61$

Draft Horse females, birth to adult $= .0794 \text{ Chest Girth}, + 2.85$

Shetland Pony males, birth to adult $= .0670 \ (\text{Chest Girth} + 31.50)$

Shetland Pony females, birth to adult $= .0587 \ (\text{Chest Girth} + 38.59)$

Light Horse females, birth to adult $= .890 \text{ Male Cannon} + 0.43$

Draft Horse females, birth to adult $= .886 \text{ Male Cannon} + 0.51$

Shetland Pony females, birth to adult $= .880 \text{ Male Cannon} + 0.40$

That is, the *sex difference* in Cannon Girth increases approximately with absolute size (body weight) :

Average Draft female $= .9272$ male (9.16 : 9.88)

Average Quarter Horse female $= .9429$ male (7.40 : 7.85)

Average Thoroughbred female $= .9436$ male (7.64 : 8.10)

Average Standardbred female $= .9470$ male (7.50 : 7.92)

Average Arab female $= .9680$ male (7.12 : 7.36)

girth measurements in addition to that of the chest are used. This formula, for adult male horses of all breeds, is:

$$\text{Body Weight, pounds} = (\text{Neck}^2 + \text{Chest}^2 + \text{Forearm}^2 + \text{Gaskin}^2 + \text{Hind Cannon}^2 \times \text{Trunk Length}) \div 392.7 \ (\pm 2\%)$$

---

* For estimating the weights of Shetlands, see directions on p. 38.

For adult female horses, or mares, the same formula is used with the divisor 395.7 (± 1.6%).

For newborn foals of both sexes the divisor is 425.0 (± 2%).

The important measure of Trunk Length varies in its relation to Withers Height according to breed. However, when Trunk Length is related to *Chest Girth,* the typical ratio obtained is a remarkably uniform 88 percent in both sexes, ranging only from 86 to 90 percent. Thus, if the Chest Girth were 70 inches, the expected Trunk Length would be 61.6 inches, with a probable range of between 60.2 and 63.0 inches. These relationships pertain to adult horses and ponies of all breeds. In newborn foals there appears to be a wider range, from about 86 percent (draft) to 96 percent (Arab), with an average ratio of 90. But the Trunk Length/ Chest Girth ratio certainly exemplifies the basic dimensional design of the equine body.

Another consistent ratio is that of Chest Depth (sternum to withers) in relation to Trunk Length, which averages close to 45 percent in adult horses of both sexes and ranges from 43.5 in Quarter Horse mares to 45.5 in Arabs. In newborns, again there is greater variation, the ratio being about 44 in Draft foals and a high 48 percent in Arabs.

An indefinite number of other correlations could be presented if any useful purpose were thereby served. Our object here is simply to indicate how most linear measurements of the body are related linearly to each other and to the square root or cube root of the body weight. From these relationships, along with the information imparted in Part I, the typical or optimum size and bodily proportions of any breed of horse of either sex and at any age may be estimated.

2. *Distribution of the Body Weight to the Fore and Hind Quarters*

As is well known to horsemen, the amount of weight borne by the fore limbs of a horse is greater than that borne by the hind limbs. However, it is not so well known that the difference in this respect is greater in ponies and small horses than in large ones. Goubaux and Barrier present a table in which fourteen horses and one donkey are listed according to height, total body weight, and weight on the fore and the hind limbs.* When these figures are plotted on a graph, it develops that the weight on the fore limbs is equal to .4961 x total body weight, + 77 pounds. This correlating formula yields the following poundages and ratios:

---

* This table is republished in Axe's encyclopedia on the horse, vol. 1, p. 93.

Body Weight, lbs.	Weight on Fore Limbs, lbs.	Percent of Total Weight
300	226	75.3
500	325	65.0
1000	573	57.3
1500	821	54.7
2000	1069	53.4

Thus in an average-sized saddle horse, standing 62 inches at the withers and weighing 1040 pounds, the amount of weight on the fore quarters would be expected to be approximately .4961 x 1040 + 77, or 593 pounds. This would amount to 593/1040, or 57 percent, of the total body weight. From this formula a plus-or-minus deviation of at least 11 percent must be allowed to accommodate the peculiarities of individual horses. And, of course, in order for the formula to apply, the horse must stand evenly on all four feet, on a level surface. Once the forequarters are raised (as in the take-off in jumping), more of the horse's weight is thrown onto the hind limbs, and vice versa. Even for the horse's head to be raised upward and backward, or lowered and drawn toward the chest, may alter the amount of weight on the fore limbs by as much as four percent.

Since a horse's center of gravity is located nearer the front of his body than the hind, it would appear that when an average-sized rider is in the saddle—which is placed *behind* the animal's center of gravity—the respective amounts of weight on the horse's fore and hind limbs should be nearly equalized. However, I know of no tests that have been made to determine this.

In addition to Goubaux and Barrier's table, S. H. Chubb, formerly a preparator at the American Museum of Natural History,* in 1920 measured Man o' War and in 1930 Gallant Fox, both as three-year-old racing Thoroughbreds. Chubb, among other measurements, gives these figures: total body weight, Man o' War 1150 pounds and Gallant Fox 1125 pounds; weight on fore quarters, Man o' War 675 pounds and Gallant Fox 645 pounds. These figures yield forequarters/body weight ratios of 58.7 percent in Man o' War (a relatively high ratio) and 57.3 percent in Gallant Fox, the latter corresponding very closely with the formula derived from the weights given by Goubaux and Barrier.

---

* In *Natural History*, vol. 31, no. 3, May–June 1931, p. 326.

3. *Weights of the Fore and Hind Limbs in the Horse*

Doubtless the typical weights of the various segments of a horse's body—head, neck, trunk, and limbs—from birth to maturity in both sexes could be established for the various breeds provided some initial data on the subject were available. Although it is not likely that such information would be of interest to the practical breeder, it might be of help to such students of hippology as are interested in the *body mechanics* of the horse. The following estimations I made on the basis of the external *linear* measurements of the limbs in the horse.

It would appear that in the newborn foal the four legs weigh together about 29 percent of the total body weight, in a yearling about 12 percent, and in the adult horse about 10 percent. Hence in a typical newborn Thoroughbred the four legs would weigh about 33 pounds, in a year-old foal about 76 pounds, and in a five-year-old stallion about 117 pounds. Conversely, the small chest and short trunk of the newborn foal yield a trunk weight that is only about 71 percent of the total body weight, while in the adult the ratio is about 90 percent. That is, in the newborn foal the limbs are both longer and heavier in relation to the small body than in the adult horse, where these proportions are reversed. Doubtless that is one reason why the foal—especially in wild forms of Equidae, such as zebras—within only a few hours after birth has enough strength in its limbs to keep up with its mother and follow the herd.

It would appear that the hind limbs weigh, on the average, in the adult horse about $1\frac{1}{6}$ times the weight of the fore limbs, and in the newborn foal about 1.15 times as much. That is, the weight of the hind limbs in relation to the fore limbs would seem to be fairly constant throughout the period of growth. However, much more information is needed in this connection, preferably from the actual weighing of the severed limbs of dead horses and foals. The length of the fore limb should be measured from the bottom of the chest (sternum) to the sole of the hoof, and the hind limb similarly, from the bottom of the barrel just in front of the lower thigh.

4. *Surface Area of the Body in the Horse*

Here is another subject, possibly of interest only to academicians, yet having a connection with the geometry of the equine body and its relationship to muscle cross section, basal metabolism, and feed requirements. A simple formula for determining the approximate surface area in light horses is: surface area in square meters $=^3\sqrt{\text{body weight, kg.}^2}$ x .0821.* In English measure this becomes: surface area in square feet $=^3\sqrt{\text{body weight, lbs.}^2}$ x .5233. The latter formula yields these areas: in a 100-pound foal, 11.27 sq. ft.; in a 600-pound pony, 37.23 sq. ft.; in a 1000-pound saddle horse, 52.33 sq. ft. For draft horses, and Shetland ponies of draft-horse build, the figures yielded by the foregoing formula should be reduced x 0.931. For a 400-pound Shetland this gives 26.45 sq. ft., and for a 2000-pound draft horse 77.34 sq. ft.

5. Gestation Length in Various Breeds of the Horse

The following figures for the length of gestation are given by K. Alminas (1939) :

	Days	
Male foals, average	337.4	(303–369)
Female foals, average	335.7	(300–365)
Arab or Thoroughbred males	339.5	
Arab or Thoroughbred females	337	
Belgian draft males	333.8	
Belgian draft females	331.5	

For Shetland ponies, J. E. Flade (1959) gives 333 days for males and 332.9 days for females.

An investigator named Tessier reported a range of 287-419 days, or about 9½ to nearly 14 months, among 582 mares, which indicates that even this extensive range does not represent the limit of variability. Foals born in the spring may average as much as 14 days longer than those born in the fall.

Young mares can breed as early as their second year, although three years is to be preferred as a minimum age. Mature mares may reproduce until an advanced age; 14 to 16 years is common, while an extreme instance of 39 years (!) was reported a few years ago in a Suffolk draft mare.

---

* In: Univ. Missouri Agric. Exper. Sta. Res. Bulletin 208, p. 8 (1934).

6. *Some Figures on the Physiology of the Horse**

The average body (rectal) temperature of the horse is 100.5 degrees F., with a normal range of from 99 to 100.8 degrees. In a day-old colt it was found to be 102¾ degrees. In donkeys the average is about 101 degrees, and in mules 102 degrees. Among farm animals, horses are nearest to man's average of 98.6 degrees. Cattle average 101.5, sheep 102.3, swine 102.6, and goats 103.8 degrees.

The pulse rate and the breathing rate vary inversely according to the size of the animal. In the horse the normal range in the pulse is from 32 to 44 beats per minute, and in the respiratory rate from 8 to 16 breaths per minute, the latter increasing rapidly with exercise. For comparison, man averages 72 heartbeats per minute, an elephant only 30 beats, and a shrew 1000 beats. In fact, someone has figured out that the typical number of heartbeats per minute in mammals is equal to the body weight in pounds raised to the minus 0.27 power (i.e., heartbeats = body weight, lbs.$^{-0.27}$).

The lung capacity in a 1000-pound horse is about 2500 cubic inches; and each ordinary inhalation accounts for about 250 cubic inches. These figures are nearly ten times those of the average man, weighing 155 pounds. The weight of the lungs in a horse weighing 1320 pounds was found to be 11.9 pounds. In proportion to body weight, this is over forty percent greater than in cattle, in which the demand for endurance and great breathing power is much less.

In keeping with the need in the horse for prolonged running ability and exceptional cardiovascular efficiency, the weight of the heart in relation to body weight is over *twice* as great as in cattle. The weight of the heart in a 1320-pound horse was found to be 9.4 pounds (or about $\frac{1}{14}$ of the body weight), while in a 1540-pound steer it was only 5.1 pounds ($\frac{1}{30}$ of the body weight).

Among various other animals that were tested, the horse has been found to be the champion sweater. During a 24-hour period while at rest, an average-sized horse lost 6.4 pounds of perspiration; in trotting, this was increased to 14 pounds; and in one horse after a trail test the loss amounted to 34 (!) pounds. Donkeys and mules sweat less profusely than do horses.

The muscular and skeletal development of the horse is a subject of special interest, since it is directly correlated with the ability of the animal to perform in its particular field (e.g., running or load pulling). The following particulars on muscular and skeletal

---

* Most of these data have been abstracted from my earlier book, *The Empire of Equus.*

anatomy in the horse have been taken from various earlier studies, mostly made by German zoologists, who laboriously derived the information through careful dissections and measurings.

In a horse of the German Trakehner or cavalry breed, weighing 456 kg (1005 pounds), the separate weights of various parts and organs were as follows: bones of each forelimb (fresh, not dried), about $12\frac{2}{3}$ pounds; each hind limb, 16.3 pounds; trunk, 56.4 pounds; skull (including lower jaw), $18\frac{1}{3}$ pounds; total $132\frac{2}{3}$ pounds, or 13.2 percent of the living body weight. In the same (freshly dissected) condition, the bones of an average man constitute about 15 percent of his body weight, and of a well-trained athlete about $13\frac{3}{4}$ percent. After the bones, either of a horse or a man, have become thoroughly dry, they weigh only about 57 or 58 percent as much as when fresh. Hence in the above 1000-pound horse, the *dry* bones would weigh about $76\frac{1}{2}$ pounds, or 7.6 percent of the body weight. That a general correspondence between skeleton weight and body weight exists among mammals of various sizes and types is indicated by the fact that in the domestic cat the relative fresh weight of the bones is about 13.6 percent of the body weight; in dogs (well nourished but not fat) 14–15 percent; in beef cattle 14.7 percent; and in small bats 14.6 percent. In animals that normally carry a great deal of fat the ratio is correspondingly lower. Thus in guinea pigs it averages only 8.8 percent, in mice 8.4 percent, in rabbits 8.1–9.2 percent, and in fat swine only 6.3–6.7 percent. Surprisingly, in the elephant, whose ponderous bulk consists mainly of bone and muscle rather than fat, the weight of the skeleton rises to over 18 percent of the living body weight.

The relative weights of the muscles of the horse and other animals show a similar correlation with that of the entire body. In both the horse and in man the muscles comprise, on the average, about 48 percent of the body weight. In domestic sheep the ratio is about 40 percent; in albino mice, 45.4 percent; and in the elephant about 51.5 percent. The amount of blood in a 1000-pound horse is about 66 pounds, or approximately $\frac{1}{15}$ of the body weight. In the above-cited 1000-pound Trakehner horse the head weighed about $39\frac{1}{2}$ pounds, the hide (without hoofs) $33\frac{1}{3}$ pounds, the four hoofs $6\frac{1}{2}$ pounds, the heart 7.2 pounds, and the brain 25.36 ounces or a little over $1\frac{1}{2}$ pounds. The latter weight is quite typical for a horse of average size. In a Shetland pony weighing 385 pounds the brain weight was 18.24 ounces or about $1\frac{1}{7}$ pounds; and in a large draft horse weighing about 2100 pounds it was 31 ounces or $1\frac{15}{16}$ pounds.

The specific gravity of a horse's body averages about 1.026. This is practically identical to the same ratio in the cat, dog, deer, and other four-footed mammals in which the measurement has been taken. In man the SG is generally given as 1.040, the higher figure possibly being due to his relatively heavier limbs and smaller chest and abdominal cavities.

7. *Growth and Attrition of the Teeth in the Horse*

Information on judging the age of a horse by its lower incisor teeth is to be found in practically every book in which the physical structure of this animal is touched upon. In contrast, hardly ever is information presented concerning the grinding or cheek teeth, and how *they* wear with age. Some writers have stated that a horse's teeth grow continuously throughout the lifetime of the animal. This is not the case. Essentially, the full growth of the teeth is attained between the ages of four and five years, after which time the teeth are *shortened* (from root to crown) as they are worn down from use. And as each tooth thus shortens, the space its root formerly occupied in the jaw is filled by a growth of new bone beyond the root. Whatever small amount of growth may take place after the age of five years is nullified by the greater rate at which attrition of the crowns of the teeth takes place.

During my extensive investigations of the skulls, skeletons, and teeth of Equidae of all kinds, a correlating of the size of the cheek (molar and premolar) teeth with age revealed that the wearing-down, or attrition, of the grinding surfaces takes place, not at a uniform rate, but in relation to the *logarithms* of the ages. This logarithmic (percentile) rate of wear is confirmed by the reduction in the fore-and-aft length of the six-tooth row that takes place as various levels of the teeth come into use. This rate of wear is shown diagrammatically in Figure 20, where each doubling of the age results in the same amount of shortening of the teeth. For example, the amount of shortening is the same between 5 and 10 years as between 10 and 20 years, or between 20 and 40 years. At the latter age—if the horse lives that long—the teeth will be worn down to the roots. In this connection it should be most interesting to examine the teeth of certain horses that lived assertedly to the age of fifty years or more!

All the foregoing would appear to be directly connected with the feeding habits of the individual horse and consequently the *rate*

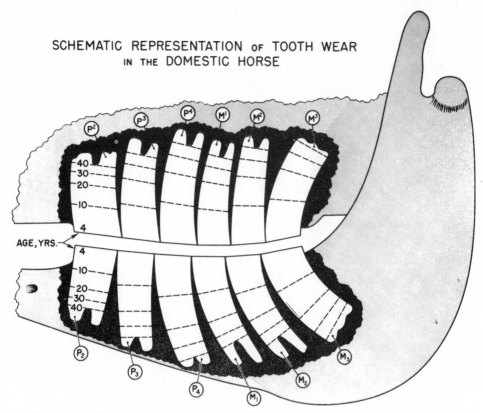

Fig. 20. *Schematic sections taken through the upper and the lower
P2-M3 series, showing the relative anteroposterior diameters and crown lengths of the cheek teeth in the domestic
horse at 4 years of age. The broken lines on the teeth indicate the respective age levels from 10 to 40 years. It is
assumed that by the age of 40 the horse's teeth, under normal living conditions, will be worn down to the roots.
Note that the teeth are abraded not at a uniform rate,
but in ratio to the logarithms of the ages, the same as
the diminution of the anteroposterior diameters of the
teeth.*

at which its teeth are worn down. In wild, desert-living Equidae,
such as the Mongolian (Przevalsky's) horse, the kulan, and the on-
ager, it would seem that the teeth must wear down faster on
account of the constant presence of sand in the herbage eaten. How-

ever, to my knowledge, no study has as yet been made of the relationship of the food consumed to the degree of wear in a horse's teeth. This may be because generally there is no great incentive to prolong a horse's life beyond the stage at which it can still produce offspring. In horse-and-buggy days there *was* such incentive—which perhaps is why most instances of extreme longevity in the horse are, or were, of light-carriage individuals (which were given the best of care) .

8. *Eruption Time of the Teeth in the Horse*

The following table, which is based on general experience, gives the usual eruption time of the teeth in the domestic horse. The abbreviated designations used are as follows: di = deciduous incisors; dc = deciduous canines (usually in male foals only) ; dm = deciduous premolars; I1 = first permanent incisors; and so on, for the remainder of the permanent teeth. In designating an upper tooth, the second number in the abbreviation is shown *above* the initial denoting the tooth; and in a lower tooth, *below* the initial. For example, P^1 designates the *upper* first premolar, and M_3 the *lower* third or last molar.

The cheek, or grinding, teeth of the horse normally consist of three premolars and three molars on each side of each jaw. The lower teeth, although forming a row of grinding surfaces as long (from front to back) as the upper row, are considerably narrower and less square than the upper teeth. Both the upper and the lower rows of these high-crowned (hypsodont) teeth grow until 4 or 5 years of age, at which time the incisors and canines reach their full length. The cheek teeth continue to grow for a somewhat longer period, so long as their roots remain open. As the teeth grow in length, they are "pushed" into chewing position by the steady growth of *bone* beneath them. However, the wear on the teeth is greater than their growth, so that with age the crowns of the teeth gradually become shorter.

9. *Some Popular Fallacies Regarding Equine Body Conformation*

In my earlier book, *The Empire of Equus,** I have gone into detail in discussing the many inaccuracies or unwarranted generalities of size and proportion commonly attributed to horses in general

---

* Cranbury, New Jersey: A. S. Barnes & Company.

TABLE 30
Eruption Time of the Teeth in the Horse

Deciduous dentition (milk teeth)		Time of eruption
1st incisor	(di 1/1)	birth to 7 days
2nd "	(di 2/2)	4 to 6 weeks
3rd "	(di 3/3)	8 to 10 months
Canine	(dc)	(absorbed without eruption)
1st premolar	(dm 2/2)	birth to 2 weeks
2nd "	(dm 3/3)	" " " "
3rd "	(dm 4/4)	" " " "

	Permanent dentition		Time of eruption
	1st incisor	(I1)	$2\frac{1}{4}$–$2\frac{3}{4}$ years
	2nd "	(I2)	$3\frac{1}{2}$–4 years
	3rd "	(I3)	$4\frac{1}{2}$–5 years
	Canine	(C)	4–5 years (rarely in female)
	1st premolar	(P1)	
	(or "wolf tooth")		5–6 months (frequently
Order of			absorbed without eruption)
appearance			
3	2nd premolar	(P2)	$2\frac{1}{4}$–$2\frac{3}{4}$ years
4	3rd "	(P3)	$2\frac{1}{2}$–$3\frac{1}{2}$ years
6	4th "	(P4)	$3\frac{3}{4}$–$4\frac{1}{4}$ years
1	1st molar	(M1)	10–12 months
2	2nd "	(M2)	20–26 months
5	3rd "	(M3)	$3\frac{1}{4}$–$4\frac{1}{4}$ years

These eruption-times are for *upper* teeth;
the lower teeth may erupt about 6 months earlier.

and to Arabs in particular (e.g., "sloping shoulders," "broad fore-head," "short back," "flat bone," "sloping pasterns," "dense bone," etc.). These and other real or imagined equine variations have been cited by writers who evidently wish their particular breed of horse to be regarded as distinctive and superior. However, a brief study of zoology, and particularly a measuring of the skulls and limb bones of horses of various breeds, would temper such assertions.

For example, the slope of the shoulders in a horse is dependent upon the lengths of the scapula and the humerus as compared with the vertical depth of the chest (from withers to sternum). This, in horses of the light or saddle type, results in the long axis of the scapula being inclined from the horizontal about 58 degrees on the average, with a normal range of from 51 to 65 degrees. In heavy draft horses the average inclination is steeper, being about 62 degrees. These are typical inclinations common to Equidae in general; and while there are, of course, individual deviations from the average, it is doubtful whether any entire breed of saddle horses can be said to have a slope of shoulder more conducive to easy riding than occurs in other breeds. The same is true of the slope of the pasterns; for while the much-prescribed "ideal" angle of 45 degrees in the front pasterns *does* sometimes occur, it is a radical departure from the *typical* inclination which in light horses averages 63 degrees from the horizontal and in draft horses 66 degrees. In the hind pasterns these inclinations are 2 or 3 degrees steeper, averaging 66 degrees and 68 degrees, respectively.

Evidently a relatively broad forehead in a horse is considered advantageous by reason of the eyes being placed farther to the sides and so enabling the animal to have a wider range of vision. But the width of the forehead in relation to head length is more a function of general body size than of breed. In the skull—which affords far greater accuracy of measurement than does the living head—small ponies, such as the Shetland, Skyros, and Lofoten (Norse) breeds, have generally the broadest foreheads as determined by the Cephalic Index, which is the ratio of the forehead or frontal width divided by the basilar length of the skull. In these ponies the index ranges from 44 to 46, although two Barb skulls showed an average of 44.2. Arabs and Thoroughbreds range from 42 to 43, and draft horses from 41 to 43. However, the smallest indices appear in the skulls of Lipizzan and Kladruber horses, with 40.5 and 40.4 respectively. So, here is another equine characteristic the supposed value of which it would be difficult to demonstrate. Of course, no one wants a horse with a freakishly narrow head; but the difference between what may be called narrow and broad, respectively, in a horse of average size amounts only to about a half-inch in ten inches.

As to the length of the back, it is due mainly to the number of lumbar vertebrae, this number being 5 or 6 on the average in Arab horses and nearly always 6 in most other breeds, although 5 sometimes occurs in Barbs and crossbreeds derived therefrom, such as Andalusians. Thus the so-called "short" back of the Arab is typically shorter by not more than the length of one lumbar vertebra (or about 1.9 inches). As to the shape and density of the bones—specifically, the cannon or metapodial bones—there is simply no difference, apart from size, between those of various breeds. As to *speed* in running, the cannon bones should be relatively *long*, not short;

and in the Arab, Thoroughbred, and numerous other light breeds, these bones *are* relatively long. It is both amusing and disconcerting to one familiar with these facts to note the ways in which most artists proportion their horses—with overly short and pointed ears, too-small heads, swan-like necks, too-long croups, and cannons too short in relation to forearms and to withers height. Every one of these equine characteristics is *measurable,* and since there are abundant studies in which they *have* been measured, there is little or no excuse for supposedly well-informed writers on horses to keep on perpetrating the fictions.

Appendix: Part II

Formulas for Daily Requirements of Nutrients

THE FOLLOWING FORMULAS, WHICH HAVE BEEN USED HEREIN TO EX-press quantitatively the principal nutrients required by horses, are of necessity more complex than those presented earlier for body measurements. One reason for this is that most of the nutrition elements are required in amounts that correspond with what is known as Metabolic Body Size, and this measure is a derivative of body weight raised to the $3/4$ power (i.e., $BW^{0.75}$). In growing foals the exponent is 0.52 (i.e., $BW^{0.52}$).

The measurement procedure favored herein is the Total Digestible Nutrients (TDN) System, mainly because the prescribed quantities are expressed primarily in *pounds* (as is done by Morrison and his followers) rather than in Metric units. This system, it would appear, is simpler to understand and to calculate than is the Digestible Energy (DE) System, as presented in the National Academy of Sciences Bulletin No. 6, and elsewhere. That the Metric system will duly become the standard in the United States, as it is elsewhere in the world with the exception of the British Empire, is a foregone conclusion. However, the transition will not take place overnight, and meanwhile the great majority of Americans will continue to express quantities in feet and inches, pounds, gallons, and other familiar units.

It may be added that one reason for the presentation here of formulas that specify the amounts of nutrients to be used is because in other publications on the subject such formulas are conspicuous by their absence. While the tables presented in such publications are manifestly the result of formulated quantities, the amounts specified in the tables are in places highly erratic, with small jumps between some successive body weights and large jumps between others, even those immediately adjoining. The following formulas avoid such unwarranted undulations, and prescribe quantities that follow consistent-

177

ly the basis (usually the Metabolic Body Size) from which they are derived.

The following formulas apply specifically to the quantities of nutrients recommended for *mature* horses, as listed in Tables 22 and 23. In these formulas, Metabolic Body Size, as listed in Table 21 is expressed as Relative Body Weight (RBW).

Daily Requirements of Nutrients for Mature Horses, Ponies, Donkeys, and Mules

Dry Matter, lbs., horses at rest $= 11.30$ $RBW^{0.75}$
"   "   "   horses at light work $= 14.69$ "
"   "   "   horses at medium work $= 16.95$ "
"   "   "   horses at heavy work $= 19.77$ "
"   "   "   mares, last quarter of pregnancy $= 15.37$ "
"   "   "   mares, nursing foals $= 22.26$ "

Digestible Protein, lbs., horses at rest $= 0.665$ $RBW^{0.75}$
"   "   "   horses at light work $= 0.815$ "
"   "   "   horses at medium work $= 0.914$ "
"   "   "   horses at heavy work $= 1.039$ "
"   "   "   mares, last quarter of pregnancy $= 0.900$ "
"   "   "   mares, nursing foals $= 2.000$ "

Total Digestible Nutrients (TDN), lbs., horses at rest $= 7.83$ $RBW^{0.75}$; $= .693$ Dry Matter, lbs.
"   "   "   "   "   horses at light work $= 10.52$ " ; $= .716$ "   "   "
"   "   "   "   "   horses at medium work $= 12.32$ " ; $= .727$ "   "   "
"   "   "   "   "   horses at heavy work $= 14.57$ " ; $= .737$ "   "   "
"   "   "   "   "   mares, last quarter of preg. $= 10.56$ " ; $= .717$ "   "   "
"   "   "   "   "   mares, nursing foals $= 17.45$ " ; $= .756$ "   "   "

(TDN is generally considered to equal: Digestible Crude Protein plus
Digestible Carbohydrates plus 2.25 Digestible Crude Fat)

Calcium, grams, horses both idle and working $= 13.80$ $RBW^{0.75}$; lbs. $= 0.0304$ $RBW^{0.75}$
"   "   mares, last quarter of pregnancy $= 15.59$ " ; lbs. $= 0.0344$ "
"   "   mares, nursing foals $= 24.43$ " ; lbs. $= 0.0538$ "

Phosphorus, grams, horses both idle and working $= 15.18$ $RBW^{0.75}$; lbs. $= 0.0335$ $RBW^{0.75}$
"   "   mares, last quarter of pregnancy $= 15.59$ " ; lbs. $= 0.0344$ "
"   "   mares, nursing foals $= 22.20$ " ; lbs. $= 0.0489$ "

Vitamin A (carotene), Mg., horses both idle and working $= .05$ (1/20) Bodyweight, lbs.
"   "   "   "   mares, last quarter of pregnancy $= .06$ Bodyweight, lbs.
"   "   "   "   mares, nursing foals $= .06$ Bodyweight, lbs.

Net Energy, therms, horses at rest $= 6.26$ $RBW^{0.75}$; $= .8000$ Total Digestible Nutrients
"   "   "   horses at light work $= 8.83$ $RBW^{0.75}$; $= .8233$ "   "   "
"   "   "   horses at medium work $= 10.55$ $RBW^{0.75}$; $= .8467$ "   "   "
"   "   "   horses at heavy work $= 12.69$ $RBW^{0.75}$; $= .8710$ "   "   "
"   "   "   mares, last quarter of pregnancy $= 8.87$ $RBW^{0.75}$; $= .8400$ "   "   "
"   "   "   mares, nursing foals $= 15.17$ $RBW^{0.75}$; $= .8693$ "   "   "

Formulas for the nutrient requirements of foals, in comparison with those for mature horses, are generally more complex, on account of the requirements for growth having to be added to those for maintenance. Some authors have attempted to prescribe, for example, the daily DE requirements on the basis of the daily gain in body weight in the foal, but the recommendations are largely invalidated by being based on excessive weight gains and on the unwarranted generalization that mature body weight is attained at only 3½ years of age. So-called maturity may, indeed, be approximated at the latter age in very small ponies and donkeys, but certainly not in normally developing horses and mules.

It should be noted that the Metabolic Body Size here assumed for growing foals is $BW^{0.52}$, rather than $BW^{0.75}$ as adopted for mature horses. The $BW^{0.52}$ has been derived from data on growing colts presented by Brody, et al (Bull. 368, 1943). Ratios of Relative Body Weight$^{0.52}$ are given here in Table 21.

The following recommended quantities are based on the rates of growth listed in Part I herein, and as far as possible make use of formulas similar to those which have just been presented for mature horses. From these formulas have been derived the figures listed in Tables 24 and 25.

Daily Requirements of Nutrients for Growing Foals
(of horses, ponies, donkeys, and mules)

Dry Matter, lbs., for				200	lbs, mature body weight			$= 14.06$	$RBW^{0.52}$	
"	"	"	"	400	"	"	"	"	$= 14.76$	"
"	"	"	"	600	"	"	"	"	$= 15.50$	"
"	"	"	"	800	"	"	"	"	$= 16.28$	"
"	"	"	"	1000	"	"	"	"	$= 17.10$	"
"	"	"	"	1200	"	"	"	"	$= 17.95$	"
"	"	"	"	1400	"	"	"	"	$= 18.84$	"
"	"	"	"	1600	"	"	"	"	$= 19.78$	"
"	"	"	"	1800	"	"	"	"	$= 20.77$	"
"	"	"	"	2000	"	"	"	"	$= 21.80$	"

In the foregoing prescriptions, the Dry Matter for each 200-pound BW increase is 5 percent greater. The quantitative Dry Matter requirements for growing foals are approximately *halfway between* those for mature horses of the same BW at *medium* and at *heavy* work, respectively.

Digestible Protein, lbs., for				200	lbs. mature BW			$= 0.61 - (.741$	Rel. $BW^{0.52})$	
"	"	"	"	400	"	"	"	$= 0.87 - (.755$	"	")
"	"	"	"	600	"	"	"	$= 1.08 - (.758$	"	")
"	"	"	"	800	"	"	"	$= 1.26 - (.761$	"	")
"	"	"	"	1000	"	"	"	$= 1.41 - (.764$	"	")
"	"	"	"	1200	"	"	"	$= 1.55 - (.765$	"	")
"	"	"	"	1400	"	"	"	$= 1.68 - (.766$	"	")
"	"	"	"	1600	"	"	"	$= 1.80 - (.767$	"	")
"	"	"	"	1800	"	"	"	$= 1.91 - (.768$	"	")
"	"	"	"	2000	"	"	"	$= 2.01 - (.769$	"	")

Digestible Protein requirements may be derived also from Dry Matter, as follows:

Digestible Protein, lbs., for 200 lbs. mature BW = 0.61 — .0530 Dry Matter, lbs.
  " " " " 400 " " " = 0.87 — .0510 " " "
  " " " " 600 " " " = 1.08 — .0490 " " "
  " " " " 800 " " " = 1.26 — .0470 " " "
  " " " " 1000 " " " = 1.41 — .0450 " " "
  " " " " 1200 " " " = 1.55 — .0430 " " "
  " " " " 1400 " " " = 1.68 — .0410 " " "
  " " " " 1600 " " " = 1.80 — .0390 " " "
  " " " " 1800 " " " = 1.91 — .0370 " " "
  " " " " 2000 " " " = 2.01 — .0350 " " "

Total Digestible Nutrients (TDN), lbs. = 0.625 Dry Matter, lbs.

$$\text{Calcium, grams} = (18 \times \text{Relative Mature BW}^{0.75}) \times \left(2.00 - \frac{\text{BW, foal}}{\text{BW, mature}}\right)$$

$$\text{Phosphorus, grams} = (15 \times \text{Relative Mature BW}^{0.75}) \times \left(1.48 - 0.6\,\frac{\text{BW, foal}}{\text{BW, mature}}\right)$$

Vitamin A (carotene), Mg. = 0.06 Body Weight, lbs.

Net Energy, therms = 0.531 Dry Matter, lbs.; =
  0.850 Total Digestible Nutrients, lbs.

While all the foregoing formulas are expressed quantitatively in Tables 22-25, they are listed here for the possible use of readers who wish to check feed quantities for certain specific body weights, and as a documentation of the nutritional procedures recommended in this book.

Bibliography

PART I: *External Body Measurements of Horses (from chapter 1: Abstracts of Previous Studies)*

Note: In all papers in our list of Abstracts where the source was not recorded, the author is omitted from the Bibliography.

Afanassieff, S. "Der Untersuchung des Exterieurs, der Wachstumintensitat und der Korrelation zwischen Renngeschwindigkeit und Exterieur beim Traber." *Zeit. f. Zucht.* 18 (1930): 171–209.

Alminas, K. "Zuchtungsbiologie Untersuchungen am litauisch-samogitischen 'Zemaitukai' Pferd im Vergleich zu anderen Rassen." *Zeitschrift fur Tierzuchtung und Zuchtungbiologie* 43 (1939) : 273–349.

Bantoiu, C. "Messungen an Trabern und die Bourteilung der Leistungfähigkeit auf Grund der mechanischen Verhältnisse." *Tierärztliche Rundschau* 28 (1922) : 666.

Bilek, F. "Über den Einfluss des Arabischen Blutes bei Kreuzungen mit bes. Hinsicht auf das Lipizzaner Pferd." *Jahrbuch f. Wiss. und prakt. Tierzucht, IX Jahrg.,* Hanover: 1914.

Boicoianu, C. "Studien uber das belgische Pferd." *Zeit. f. Zucht.* 23 (1932): 25–54.

Boing, —. 1911 (See Zimmermann, C., 1933.)

Brandes, H. "Der Einfluss der Vollblutpferdes auf die Herausbildung der ostpreusschen Halbblutzucht in bezug auf Körperform, Stärke und Leistung." *Zeitschr. f. Tier. u. Zuchtung.* 7 (1926) : 169–217.

Brody, S. "Growth and Development, with Special Reference to Domestic Animals. IX. A Comparison of Growth Curves of Man and Other Animals." Univ. Missouri Coll. Agric. Res. Bull. 104 (1927) : pp. 4–31.

————. *Bioenergetics and Growth.* New York: Hafner Pub. Co., Inc; 1964.

Butz, Henseler, and Schottler. "Praktische Anleitung zum Messen von Pferden." *Anleitungen der Deutschen Gesellschaft fur Zuchtungskunde,* vol. 2 (1921).

Crampton, E. W. "Size in the Draft Horse." *Jour. Agric. and Hort.* 26 (1923 a.): 157–158.

———. "Rate of Growth of Draft Colts." *Jour. Agric. and Hort.* 26 (1923 b.) : 172.

Cunningham, K., and Fowler, S. H. "A Study of Growth and Development in the Quarter Horse." *Bull. No. 546, Louisiana State Univ. Agric. Exper. Sta.* (1961) : pp. 3–26, 5 tables, 5 figs.

Dawson, W. M. "Height, Weight, and Girth Measurements of Eight Morgan Stallions and Ten Morgan Mares." (Pers. comm. from W. M. Dawson, Animal Husbandman, U. S. Dept. Agric., October 27, 1948).

Dawson, W. M., Phillips, R. W., and Speelman, S. R. "Growth of Horses under Western Range Conditions." *Jour. Animal Science* 4 (1945) : 47–54.

Degen, K. "Variationsstatistische Körpermassuntersuchungen an 500 Oldenburger Zuchtstuten." Inaug. Diss. Vet. Hannover: 1933.

Dimitriadis. J. N. "Das Skyrospony. Ein Beitrag zum Studium der Pferde Griechenlands." *Zeitschr. f. Zuchtung,* vol. 37, part B (1937), pp. 343–85.

Dinsmore, W. *Horse and Mule Power.* Book No. 235. Horse and Mule Association of America (1939) : p. 55.

Dunn, N. K. Personal communications to the author in July 1973.

Ewart, J. C "The Rate of Growth in the Horse." *Live Stock Journal Almanac* (1901).

Feige, E. "Über die Variation der Grosse ostpreussischer Halbblutpferde." *Zeitschr. f. Tier. u. Züch.* 10 (1927) : 241–55.

Flade, J. E. *Shetlandponys. Die Neue Brehm-Bücherei.* Wittenberg: 1959, 79 pp., 45 figs., 32 tables.

Franke, H. "Untersuchungen über den Einfluss des Korperbaus auf die Schrittlange des Pferdes." *Grevesmuhlen i. Mecklenberg* (1935). 44 pp.

Goreniuc, A. *Das Schleswigsche Pferd, seine Zucht und mechanischen Verhältnisse im Vergleich zum Pinzgauer und anders.* Berlin: 1924.

Grange, E. A. A. "The External Conformation of the Horse." Michigan State Agric. Coll. Exper. Station, Bulletin 110 (1894), pp. 67–98.

Gregory, K. "Die Mässe von 280 Stuten aus Mezöhegyes, verarbeitet zu einem Beitrag fur die Beurteilungslehre des Pferdes." *Arch. f. Tiernährung und Tierzucht,* vol. 6 part 1 (1931), pp. 18–97.

Gross, H. In *Stang and Wirth's Encyclopedia: Tierheilkunde und Tierzucht,* vol. 8 (1930), p. 58.

Gutsche, —. 1914. In *Messungen und Wägungen am Pferd,* by Dr. Reinhold Schmaltz, 1922.

Hering, A. "Ein Beitrag zur Kenntnis der Jugendentwicklung des rheinish-deutschen Kaltblutpferdes. Diss. 1925, Gottingen." *Arbeiten der Deutschen Gesellschaft für Zuchtungskunde,* no. 27. Berlin: 1925.

Hervey, J. "Champion Stallions and Their Conformation." *The Harness Horse,* vol. 7, no. 7 (Dec. 10, 1941), p. 112.

Hooper, —. In an article in *The Thoroughbred Record* (July 9, 1921).

Iwersen, E. "Die Körperentwicklung des holsteininschen marschpferdes von der Geburt bis zum Abschluss des Wachstums." *Züchtungskunde* 1 (1926): 134–43.

Kolbe, W. "Das Oberlander Pferd unter besonderer . . . Normanner, Clydesdaler und Clevelandstammes." *Abhand. des Inst. fur Tier. und Molk, an der Univ. Leipzig 15* (1928): 1–87.

Kronacher, C., and Ogrizek, A. "Exterieur und Leistungsfahigkeit des Pferdes mit besonderer Berucksichtigung . . . " *Zeit. f. Zuch.* 23 (1932): 183–228.

Kruger, W. "Über Wachstumsmessungen . . . der Trakehner Warmblut und Mecklenburger Kaltblutpferd." *Zeit. f. Tier u. Zuch.* 43 (1939): 145.

Kuffner, H. "Studien über das orientalische Pferd mit besonderer Berücksuchtigung seiner Zucht in Balbona." *Arb. Lehr. f. Tierzücht an der Hoch. f. Boden in Wien* 1 (1922): 151–92.

Lesbre, F. X. "Des proportions du squelette du cheval, de l'ane et du mulet." *Bull. Soc. d'Anthrop. Lyons* 12 (1893): 125–44.

———. "Etudes hippometriques." *Jour. Med. Veterinaire et de Zootechnie* (1894), 14–24; 75–88; 150–56; 208–22.

Letard, E. "L'utilitie des mensurations pour appreier le Developpment des Animaux domestiques." *Revue de Zootechnie,* nos. 9 and 10, 1925.

Linsley, D. C. *Morgan Horses*. New York: C. M. Saxton and Co., 1857. 340 pp., illus.

Madroff, C. "Das Lippizanerpferd und seine Zucht in Europa." *Zeitschr. f. Zucht.*, vol. 33, sec. B (1935), pp. 169–84.

————. "Das Gidranpferd und seine Zucht in Europa." *Zeitschr. f. Zucht.*, vol. 36, sec. B (1936), pp. 237–53.

————. "Das Noniuspferd und seine Zucht in Europa." *Zeitschr. f. Zucht.*, vol. 36, sec. B (1936), pp. 337–57.

————. "Araberpferd und seine Zucht in den Donaulandern." *Zeitschr. f. Zucht.*, vol. 39, sec. B (1937) pp. 43–66.

Magiaru, C. *Biometrische Studie über das Lippizaner Pferd aus dem Zuchtsyndikat Sâmbata de Jos, Kreis-fagaras*. Diss., Bucharest: 1936. 70 pp.

Mieckley, —. "Wägungen und Messungen von Füllen der Trakehner Fuchsherde." *Archiv f. Wissen. u. Prak. Thier.* 20 (1894) : 320–26.

Müller, Dr. von. "Die Bedeutung der Schulter fur die Beurteilung des Pferdes." *Zeitschr. f. Veterinarkunde* 45 (1933), 175–85, 204–23, 255–64. The same, of "Brustmasse," pp. 32–50.

Nathusius, S. von. *Unterschiede Zwischen der morgenlandischen und abendländischen Pferdegruppe*. Berlin: 1891.

————. *Messungen an Hengsten, Stuten und Gebrauchspferden*. Berlin: 1905 .

Nicolescu, J. "Messungen uber die Mechanik des Hannoverschen Pferdes im Vergleich zum Vollblut und Traber." *Berliner Tierarztliche Wochenschrift,* 39 (1923), 74–77.

Ogrizek, A. "Studie über die Abstammung des Insel-Veglia (Krk) Ponys." *Arbeit. der Lehrk. f. Tier. an der Hochschule f. Bodenkultur in Wien* 2 (1923) : 73–100.

Osowicki, A. *Das Huzulenpferd. Eine züchterische Studie nach Untersuchungen in seiner Heimat*. Stuttgart: 1904.

Plischke, A. *Anatomische Körpermessungen am lebenen Vollblutpferde, ausgeführt nach der Messmethod von Schmaltz, unter vergleichsweiser Heranziehung des Trakehner Halbblutes*. Inaug. Dissert., Berlin: 1927. 40 pp.

Radescu, T. "Biometrische Untersuchungen an Vollblutpferden . . ." *Berliner Tierarztliche Wochenschrift* 39 (1923) : 490–92.

Rhoad, A. O. *A Statistical Study of Pulling Power in Relation to*

Conformation of Draft Horses. Cornell Univ.: Master's thesis, February 1928. 35 pp., 11 graphs.

Rösiö, B. *Die Bedeutung des Exterieurs und der Konstitution des Pferdes für seine Leistungsfahigkeit.* Inaug. Dissert., Berlin: 1928.

Schilke, F. *Biometrische Untersuchungen über das Wachstum der Trakehner Pferde.* Dissert., Königsberg: 1922.

Schöttler, F. *Wachstumsmessungen an Pferden. Ein Beitrag zur Entwicklung des hannoverschen Halbblutpferden.* Dissert., Schaper, Hanover: 1910.

Solanet, E. *El caballo Criollo.* Conf. Fac. Agron. Veter. Buenos Aires, 23 September, 1923.

———. *El caballo Criollo.* Buenos Aires: 155 pp., 48 figs. (An expanded version of the author's 1923 paper.)

Stang, V., and Wirth, D. In *Tierheilkunde und Tierzucht,* vol. 1. Berlin: 1926.

Stegen, H. "Die Entwicklung des hannoverschen Hengstfohlen in dem Hengstaufzuchtgestüt Hunnesruck." *Züchtskunde* 4 (1929) : 273–88.

———. "Die Entwicklung des hannoverschen Halbblutpferdes von der Geburt bis zum Abschluss des Wachstums." *Jour. f. Landwirtschaft.* 77 (1929–30) : 139–90.

Stratul, J. "Biometrische Untersuchungen an Vollblutpferden . . . " *Tierärztliche Rundschau* 28 (1922) : 665–66.

Svanberg, V. "Beitrag zur Kenntnis der Rassenmerkmale des finnischen Pferdes." *Suomal. Tiede, Toimituksia, Ann. Acad. Sci. Fennicae,* vol. 27, Ser. A, no. 4 (1928), pp. 1–87. Two pls.

Szabo, —. 1924. (See Alninas, K., 1939.)

Team, C. B. Height, Weight, Chest Girth and Cannon Girth of 25 Arab Stallions and 25 Arab Broodmares at the Kellogg Arabian Horse Ranch, Pomona, California. Personal communication of May 3, 1949.

Trowbridge, E. A., and Chittenden, D. W. *Horses—Management, Feed Consumption, and Growth Data.* III. Univ. Missouri Coll. Agric. Res. Bull. 143, 1930. pp. 149–57.

———. *Horses Grown on Limited Grain Rations.* Univ. Missouri Coll. Agric. Exper. Sta. Bull. 316, 1932. 19 pp.

Ullmann, G. "Entwicklung und Wachstum bei Shetlandponys." *Tierärztliche Rundschau,* vol. 45, no. 16 (1939), p. 312.

Unckrich, —. *Das Zweibrucker Pferd.* Inaug. Dissert. Munich: 1925.

Voltz, W. "Über die Veranderung des Exterieurs während des Wachstums beim ostpreussischen Halbblutpferde." *Landwirt. Jahrb.* 44 (1913) : 409–36.

Wiechert, F. "Messungen an ostpreussischen Kavalleriepferden und solchen mit besonderen Leistungen und die Beurteilung der Leistungsfähigkeit auf Grund der mechanischen Verhaltnisse." *Arb. der D. G. für Zucht.*, no. 34. Schaper, Hanover: 1927.

Weiss-Tessbach, A. "Studien über das Pferd des Pinzgaues." *Arb. d. Lehr. f. Tier. and der Hochschule f. Bodenkultur in Wien* 2 (1923) : 157–215.

Wilkomm, W. *Das Beberbecker Pferd. Hippologische und hippometrische Untersuchungen.* Dissert., Leipzig: 1921.

Willoughby, David P. *External Body Measurements of Three Stallions, Nine Mares, and One Foal (at 1½ Days and 38 Days, Respectively) at the El Cortijo Arabian Horse Ranch of Donald and Charles McKenna, Claremont, California.* Unpublished study. 1950.

Wöhler, H. *Variationsstatistische Untersuchungen an 200 hannoverschen Stutebuchstuten der Unterlbmarsch unter besonderer Berucksichtigung der Nelusko-Aldemann-l-Linie.* Dissert., Halle: 1927.

Zimmermann, C. "En Beitrag zur Körperentwicklung der rheinishdeutschen Kaltblutpferde in Original zuchtgebeit." *Archiv f. Tierernahrung und Tierzucht.*, vol. 8, part 4 (1933) , pp. 497–542.

Some German periodicals on horse breeding, including those cited in the foregoing Bibliography.

1. *Abhandlungen des Instituts für Tierzucht und Molkereiwesen an der Universität Leipzig.*
2. *Anleitungen der Deutschen Gesellschaft für Zuchtungskunde.* Berlin.
3. *Arbeiten der Deutschen Gesellschaft für Zuchtungskunde.* Göttingen.
4. *Arbeiten der Lehrkanzel für Tierzucht and der Hochschule für Bodenkultur in Wien.* (Vienna) .
5. *Archiv für Tiernahrung und Tierzucht.* Berlin.
6. *Archiv für Wissenschaftliche und Praktische Thierheilkunde.* Berlin.

7. *Berliner Bierarztliche Wochenschrift*. Berlin.

8. *Jahrbuch für Wissenschaftliche und Praktische Tierzuchtung*. Hanover.

9. *Journal für Landwirtschaft*. Berlin.

10. *Landwirtschaftlich Jahrbuch*. Berlin.

11. *Mitteilungen der landwirtschaftlichen Lehrkanzeln der Hochschule für Bodenkultur in Wien* (Vienna).

12. *Neue Forschungen in Tierzucht und Abstammungslehre*. Bern.

13. *Tierärztliche Rundschau*. Berlin-Friedenau.

14. *Tierheilkunde und Tierzucht*. Berlin.

15. *Zeitschrift für Gestütkunde und Pferdezucht*. Hanover.

16. *Zeitschrift für Saugetierkunde*. Leipzig.

17. *Zeitschrift für Veterinärkunde*. Berlin.

18. *Zeitschrift für Tierzüchtung und Zuchtungsbiologie, einschliessen Tierernärung*.

19. *Zeitschrift für Züchtung*. Berlin. (Same as no. 18).

20. *Züchtungskunde*. Gottingen.

NOTE: All the foregoing periodicals were in publication prior to World War II. Whether they still are, I do not know. Files of most of these periodicals are available at various agricultural colleges throughout the United States. I borrowed most of my material from the University of California Farm Library at Davis. Readers interested should consult one of the colleges of agriculture in their home state. It will be noted that most of the articles dealing with the body measurements of horses have appeared in *Zeitschrift für Zuchtung*.

PART II: *Nutrition in the Horse* (with annotations)

Amar, Jules. *The Human Motor*. New York: E. P. Dutton & Co., 1920. 470 pp., illus. A comprehensive discussion of energy and work in man, many of the conclusions of which can be applied to horses also.

Anonymous *Nutrient Requirements of Horses*. 3rd rev. ed. Washington, D.C.: National Academy of Sciences, Bull. No. 6., 1973. 33 pp., illus. A valuable reference work, although in places rather technical. Table 5 is a useful list of feeds and their composition. There are also numerous references.

Bradley, M., and Pfander, W. H. *Feeds for Light Horses* (No.

2806) , *Feeding Light Horses* (No. 2807) , and *Rations for Light Horses* (No. 2808) . Three useful, cost-free bulletins issued by the Department of Animal Husbandry, College of Agriculture, University of Missouri, Columbia, Mo.

Brody, S., Kibler, H. H., and Trowbridge, E. A. *Growth and Development.* LVIII: *Resting Energy Metabolism and Pulmonary Ventilation in Growing Horses.* Res. Bull. 368, Univ. Missouri Coll. Agric., May 1943. 14 pp., 3 tables, 4 figs. A valuable paper on oxygen consumption, ventilation rate, and heat production in relation to body weight in ponies and horses as compared with cattle.

Crampton, E. W., and Harris, L. E. *Applied Animal Nutrition.* 2nd ed. San Francisco: W. H. Freeman & Co., 1969. 753 pp., illus. A textbook on the nutrition of farm animals, with numerous author references throughout the text, and including an extensive, 272-page list of the composition of feeds.

Ensminger, M. E. *Horses and Horsemanship.* Danville, Ill.: The Interstate Printers and Publishers, Inc., 1969. 907 pp., illus. One of the most useful, comprehensive, and authoritative books on its subject. Chapter 13, on Feeding Horses, consists of 125 pages of up-to-date information.

————, ed. *Stud Manager's Handbook.* Vol. 7. Clovis, Calif.: Agriservices Foundation, 1971. 186 pp., illus. A series of monographs by various authorities, including six papers on horse nutrition and feeding.

————, collaborator. *Breeding and Raising Horses.* U.S. Dept. Agric. Handbook No. 394, 1972. 81 pp., illus. Includes nine pages of information on feeding, excerpted in part from Dr. Ensminger's book, *Horses and Horsemanship.*

Hafez, E. S. E., and Dyer, I. A., eds. *Animal Growth and Nutrition.* Philadelphia: Lea & Febiger, 1969. 402 pp., illus. A technical discussion by numerous specialists in Animal Science, Animal Husbandry, and Biology, bearing on the growth, development, and nutritional requirements of cattle, sheep, swine, and horses.

Hintz, H. F., and Schryver, H. F. *Horse Feeding.* Equine Research Program, Cornell University, Ithaca, New York. A cost-free brochure containing basic information on the subject by two prominent veterinarians.

Morrison, F. B. *Feeds and Feeding,* 22nd ed. Clinton, Iowa: The

Morrison Pub. Co., 1959. 1165 pp., illus. An indispensable guide and reference work for breeders of horses, mules, cattle, sheep, goats, swine, and poultry. The TDN system is followed. There are many tables of feed analyses, etc.

Tyznik, W. J. *Feeding Horses*. Dept. of Animal Science, Ohio State University, Columbus, Ohio. A brief resume of the essential factors in horse nutrition. Obtainable without charge.

Ullrey, D. E. *Horse Nutrition*. Animal Husbandry Dept., Michigan State Univ., East Lansing, Mich., 1971. Eleven pages of text and nine pages of tables, covering authoritatively the main aspects of the subject. Obtainable without cost.

Index